Springer Geology

Series Editors

Yuri Litvin, Institute of Experimental Mineralogy, Moscow, Russia
Abigail Jiménez-Franco, Del. Magdalena Contreras, Mexico City, Estado de México, Mexico

The book series Springer Geology comprises a broad portfolio of scientific books, aiming at researchers, students, and everyone interested in geology. The series includes peer-reviewed monographs, edited volumes, textbooks, and conference proceedings. It covers the entire research area of geology including, but not limited to, economic geology, mineral resources, historical geology, quantitative geology, structural geology, geomorphology, paleontology, and sedimentology.

More information about this series at http://www.springer.com/series/10172

Subir Sarkar · Santanu Banerjee

A Synthesis of Depositional Sequence of the Proterozoic Vindhyan Supergroup in Son Valley

A Field Guide

 Springer

Subir Sarkar
Department of Geological Sciences
Jadavpur University
Kolkata, West Bengal, India

Santanu Banerjee
Department of Earth Sciences
Indian Institute of Technology Bombay
Mumbai, Maharashtra, India

ISSN 2197-9545　　　　　　　ISSN 2197-9553　(electronic)
Springer Geology
ISBN 978-981-32-9553-7　　　ISBN 978-981-32-9551-3　(eBook)
https://doi.org/10.1007/978-981-32-9551-3

This Springer imprint is published by the registered company Springer Nature Singapore Pte Ltd.
The registered company address is: 152 Beach Road, #21-01/04 Gateway East, Singapore 189721, Singapore

*We dedicate this book to Prof. Pradip K. Bose
and Dr. Sreeradha Bose*

Preface and Acknowledgements

The Vindhyan exposures cover a large part of India and for many reasons occupy the centre stage in Indian sedimentology and deserve special attention. The Paleo- to Neoproterozoic sedimentary succession hosts the record of the biosphere, hydrosphere and atmosphere for a period that followed immediately after the great oxidation event. Only mild deformation and metamorphism ensured the record to be exceptionally well preserved, spectacular and alluring for geologists, especially sedimentologists and stratigraphers. The sedimentary succession is extraordinarily well exposed, easily accessible and a treasure house of sedimentary structures. Almost all the varieties of sedimentary structures, except those related to bioturbation, are superbly preserved in the Vindhyan. The entire gamut of lithological variations from siliciclastics to carbonates through mixed lithologies adds to its charm. Varieties of stromatolites in carbonates and very rich assemblages of microbial mat-related structures in siliciclastics, now well-known worldwide, are especially attractive. Textbook examples of seismites, tsunamiites, tempestites and tidalites make this Supergroup, especially suited for sedimentological training for budding geologists. Excellent quality of exposures allows the superb display of fluvial architectural elements and provides a facility for tracing stratigraphic bounding surfaces of different ranks. Eolian paleogeomorphic elements are preserved at selected levels. The readily workable paleogeographic shifts in terms of facies tracts make the Supergroup alluring for a budding sequence stratigrapher. It is easy to visualize paleogeographic shits, relative sea level variation and basinal evolution through time. The three-dimensional exposures of Vindhyan in many places provide the opportunity to carry out the outcrop-based sequence stratigraphy and to identify parasequences and systems tracts in certain parts.

Being superbly exposed in central India, scores of geologists worked upon this Supergroup and numerous publications in national and international journals indicate the significance of various attributes in the Vindhyan rocks. The exposure area is readily accessible, negotiable with crisscross motorable roads and habitable, with modest accommodations. Over time, new road cuttings, mine walls and water reservoirs add to new revelations. The proposed field guide containing detailed route map, geological map and plenty of coloured photographs should be useful to students, academicians and professionals for field visits in this area. We attempt to cover both the lower and upper Vindhyan rocks and both northern and southern flanks of the basin with adequate sedimentological detailing. In this book, we have presented many illustrations to show wide variations in sedimentary structures, including those related to microbial activities on siliciclastic deposits.

We felt the need of writing this field guide as we had to conduct fields for our students and also professionals, for exposing the different facets of the sedimentology and stratigraphy. One of the existing field guide books by Bhattacharyya, Chanda and Bose (1986) which covers the upper Vindhyan exclusively is out of print. The other field guide book by Kumar and Gupta (2002) focusses on Precambrian biogenic structures. The present guidebook puts more emphasis on stratigraphy, facies analysis, paleogeographic shifts, event deposits and microbial mat structures. The information presented in the book should be handy in the planning of a field trip for students and professionals, and for optimizing the duration of the field, and to maximize the learning experience. While a graduate student can learn about sedimentary structures and stratigraphy, a teacher can use the information for thorough and productive field training for students. We hope that this book serves in planning future field trips to the fascinating sedimentology of the Vindhyan.

We sincerely acknowledge and thank our colleagues and students who have supported us in preparing the book: Sabyasachi Mandal, Swagata Chakraborty, Adrita Choudhuri, Indrani Mandal, Partha Pratim Chakraborty, Snehashish Chakraborty and Megomita Das. We also thank Aninda Bose, Senior Editor, Springer Science, for initiating the book proposal and extending the deadline of submission of the book. SB thanks the Indian Institute of Technology Bombay for providing the infrastructure facilities and academic freedom for writing the book. SS thanks the Jadavpur University for providing the infrastructure support to write the book.

<table>
<tr><td>Kolkata, India</td><td style="text-align:right">Subir Sarkar</td></tr>
<tr><td>Mumbai, India</td><td style="text-align:right">Santanu Banerjee</td></tr>
</table>

References

Bhattacharyya A, Chanda SK, Bose PK (1986) Upper Vindhyans of Maihar: a field guide. Jadavpur University, Kolkata

Kumar S, Gupta S (2002) International Field Workshop on the Vindhyan Basin, Central India. Field Guide Book, Pal Soc India, Lucknow

Contents

About the Authors

Dr. Subir Sarkar is a Professor at the Department of Geological Sciences, Jadavpur University, Kolkata, India. He completed his Ph.D. at Jadavpur University in 1991. He has more than 25 years of experience in teaching and about 30 years in research. His research interests include sedimentology/ biosedimentology and basin evolution. Dr. Sarkar has published more than 100 research articles in peer-reviewed national and international journals and conference proceedings. He has also published 16 book chapters and edited four books. The Vindhyan Group is a happy hunting ground for Prof. Sarkar since 1991. Prof. Sarkar has worked in different laboratories in different countries and has accumulated vast experience about many Proterozoic formations around the world.

Dr. Santanu Banerjee is a Professor at the Department of Earth Sciences, Indian Institute of Technology Bombay since 1999. He completed M.Sc. (Applied Geology) at Allahabad University (1992) and Ph.D. at Jadavpur University (1997). He supervised several research projects on sedimentology and stratigraphy of Indian sedimentary basins. His research interests include microbial mat structures in Precambrian siliciclastics, origin of glauconite, sequence stratigraphy and petroleum geology. He has worked extensively on sedimentary facies and basin analysis in the Precambrian Vindhyan basin of central India and Meso-Cenozoic Kutch basin. Dr. Banerjee has published more than 100 research articles in peer-reviewed journals, contributed for 18 book chapters and edited 2 books. He is the country Ambassador of Society for Sedimentary Geology (SEPM). He currently serves the editorial board of Journal of Palaeogeography, Journal of Earth Systems Science and Arabian Journal of Geosciences.

Chapter 1
Geological Background

1.1 Introduction

The Vindhyan Basin is one of the largest intracratonic sedimentary basins of the world (Fig. 1.1). The basin has been studied for more than a century since the time of Oldham (1856) and Mallet (1869). Auden (1933) provided the basic data for all the subsequent studies on the Vindhyan Basin (Ahmad 1958; Banerjee 1964, 1974; Singh 1973, 1980, 1985; Chanda and Bhattacharyya 1982; Soni et al. 1987; Chakraborty and Bose 1990, 1992; Prasad and Verma 1991; Bhattacharyya and Morad 1993; Chakraborty 1993, 1995, 2001; Basumallick et al. 1996; Akhtar 1996; Bhattacharyya 1996; Sarkar et al. 1996, 1998, 2002a, b, 2005, 2006; Ram et al. 1996; Bose et al. 1997, 2001, 2015; Banerjee and Jeevankumar 2003, 2005; Ram 2005; Banerjee et al. 2005, 2006a, b, c, 2008, 2010; Paikaray et al. 2008). The sedimentation within this basin took place largely in shallow marine environments that include tidal flat (both carbonate and siliciclastic), shoreface, storm-dominated shelf, homoclinal carbonate ramp, distally steepened ramp, with fluvial and eolian intervals. The undisturbed and less metamorphosed sedimentary succession of the Vindhyan has recorded a substantial part of Mesoproterozoic and Neoproterozoic, and therefore contain crucial information about the evolution of the biosphere, atmosphere and hydrosphere of our planet. Present study on the Vindhyan Basin is not only restricted to paleogeographic interpretation but is also aimed at deciphering the sequence stratigraphic architecture of the Vindhyan Basin evolution as well as microbial mat influence on sediment depositional system (Banerjee 1997; Bose et al. 2001, 2015; Chakraborty 2004; Sarkar et al. 2005; Banerjee and Jeevankumar 2003, 2005). Recent investigations further focus on the correspondence between sedimentation, tectonics and microbial mat. The biotic remains of the Vindhyan, although fascinating, remains highly contested, particularly for its age implications. The absolute age of Vindhyan rocks, although remained doubtful till the last century, is being worked out in recent years, with a large number of radiometric dates.

© Springer Nature Singapore Pte Ltd. 2020

S. Sarkar and S. Banerjee, *A Synthesis of Depositional Sequence of the Proterozoic Vindhyan Supergroup in Son Valley*, Springer Geology, https://doi.org/10.1007/978-981-32-9551-3_1

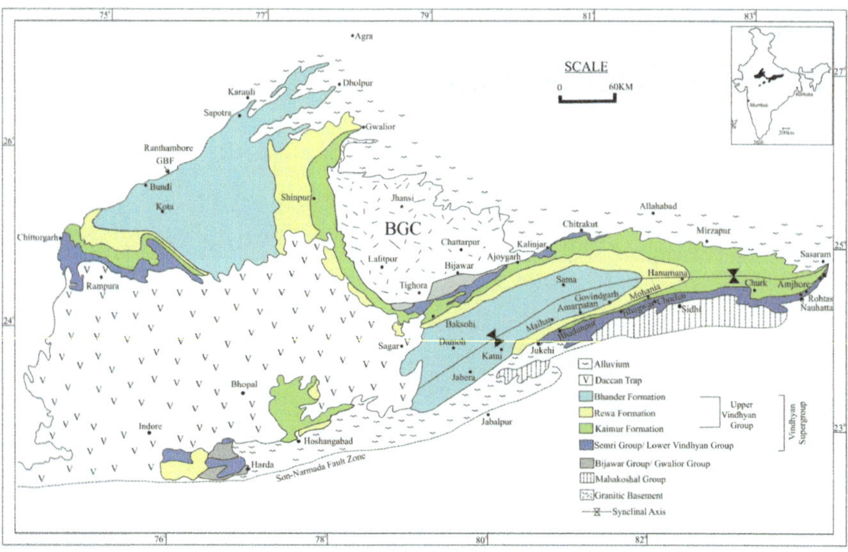

Fig. 1.1 Geological map showing outcrops of the Vindhyan Supergroup in the Son valley and Rajasthan with important locations (map of India within inset)

1.2 Outcrop Distribution

The Vindhyan Basin is the largest epicontinental Proterozoic basin in India (Fig. 1.1). The sedimentary pile is up to 4500 m thick (Ahmad 1971). The outcrops of the Vindhyan Supergroup are distributed in Son Valley (covering Madhya Pradesh, Uttar Pradesh and Bihar) and Rajasthan. These exposures cover an area of about 100,000 km^2, and an additional 70,000 km^2 is concealed under the Deccan traps (Auden 1933; Krishnan and Swaminath 1959). Archean Bundelkhand Granite Complex and the Cretaceous Deccan Traps divide Vindhyan exposures into two broad areas, viz. the Son Valley covering Bihar, Jharkhand, Uttar Pradesh, and Madhya Pradesh in central India, and the Chambal Valley covering Rajasthan and western Madhya Pradesh in Western India (Fig. 1.1). The thick alluvial cover of Yamuna, Ganga and Son rivers has covered a considerable thickness of Vindhyan sediments in central India. The stratigraphic units comprising the Vindhyan Supergroup are traceable laterally within an individual sector, but they vary in sedimentological attributes from one sector to that of the other. This field guide provides information about Vindhyan outcrops in the Son Valley.

The Vindhyan outcrop belt forms a syncline in the Son Valley area with E–W elongation and closure in the east (Fig. 1.2). The dips of the beds on two flanks converge towards the axis of the syncline. The younger rocks occur near the core of the syncline, whereas the older formations occur near the flanks of the syncline. Exposures of the Semri Group are conspicuously asymmetrical on both sides of the E–W oriented synclinal axis. The Semri exposures are discontinuous and occur as

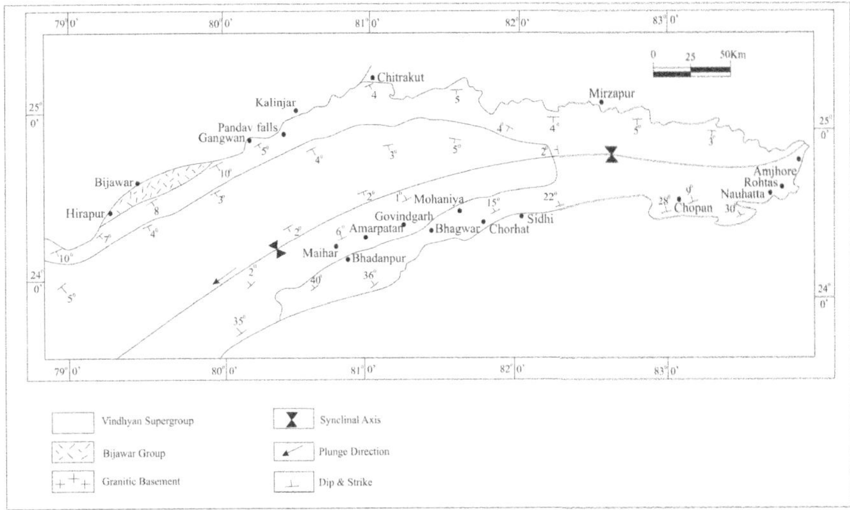

Fig. 1.2 Geological map showing outcrops of the Vindhyan Supergroup in the Son Valley with important locations

small patches to the north (Fig. 1.2). The dips of the younger formations are very gentle (up to 4°), while the dips of the oldest unit (Semri Group) generally vary between 15° and 30°. On the contrary, the outcrop is very extensive and continuous towards the southern flank of the syncline (Fig. 1.2). The Upper Vindhyan, however, is uniformly distributed on both sides of the synclinal axis. The eastward closure of the outcrop indicates the westward plunging nature of the syncline.

1.3 Lithostratigraphy of the Vindhyan Supergroup

The only major unconformity in the entire Vindhyan succession within in the Son Valley area separates the Vindhyan Supergroup into Lower and the Upper Vindhyan Groups (Figs. 1.1, and 1.2; Chanda and Bhattacharyya 1982; Bose et al. 2001, 2015). A sharp upward transition from carbonates to siliciclastics everywhere across this surface reflects a basin-wide regression of the sea (Bose et al. 2001). The base of the Supergroup is of Paleoproterozoic age (Rasmussen et al. 2002; Ray et al. 2002; Bengtson et al. 2009; Gilleaudeau et al. 2018). The upper age limit of the Supergroup, on the other hand, is more controversial: the previous general consensus was ~600 Ma (Ray et al. 2003; Ray 2006), while some recent workers suggest that it could be 900–1000 Ma (Malone et al. 2008; Gopalan et al. 2013; Venkateshwarlu and Rao 2013; Gilleaudeau et al. 2018). The Lower Vindhyan has a formal name, the Semri Group; but the Upper Vindhyan, equivalent in stratigraphic rank, does not have any.

The Semri Group comprises five formations, viz. Deoland, Kajrahat, Porcellenite, Kheinjua and Rohtas, in order of superposition (Fig. 1.3). The Deoland Formation is almost entirely sandy with local basal patches of conglomerate (Prakash and Dalela 1982; Banerjee 2010). The immediately overlying Kajrahat Limestone is divided into two parts, viz. the Arangi Shale at its base and the Kajrahat Limestone Member above (Fig. 1.3). The Arangi Shale consists mainly of dark grey and carbonaceous

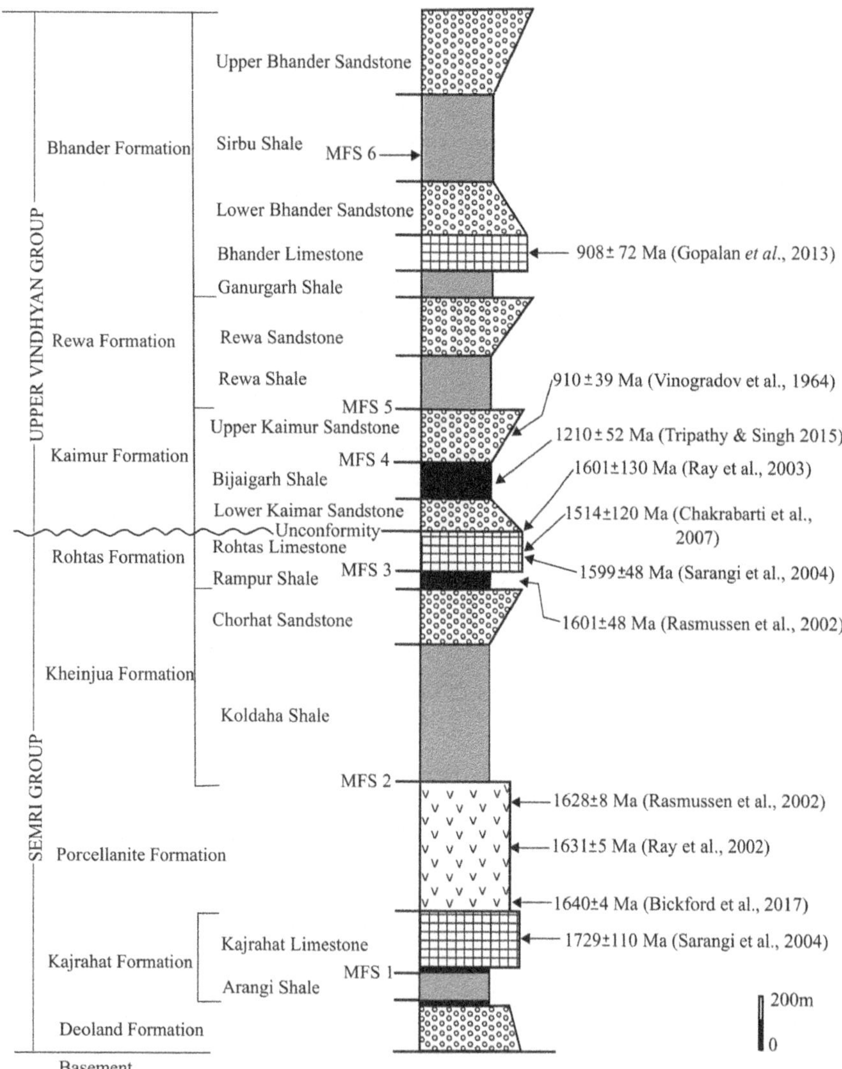

Fig. 1.3 Lithocolumn showing details of the Vindhyan stratigraphy up to the member level, lithological variations, broad depositional trends, maximum flooding surfaces and well-accepted radiometric dates

shale with some lenticular patches of carbonates, but its exposures are minimal. On the other hand, the Kajrahat Limestone consists of limestone and dolomite with the widespread development of stromatolite (Rao and Neelakantam 1978; Prakash and Dalela 1982; Banerjee et al. 2007). The sedimentary succession consisting of silicified shales, pyroclastics and volcanic tuffs, named as the Porcellanite Formation, overlies the Kajrahat Formation. The Porcellanite Formation gradationally passes over to the Kheinjua Formation that consists of shale, sandstone, limestone and patches of intraformational conglomerate (Auden 1933; Rao and Neelakantam 1978; Prakash and Dalela 1982; Bose et al. 1997, 2001; Banerjee 2000; Sarkar et al. 2006). This formation consists of two members, Koldaha Shale at the base and Chorhat Sandstone at the top. The Rohtas Formation at the top of the Semri Group consists of the Rampur Shale Member below and the Rohtas Limestone Member above (Chatterjee and Sen 1988; Banerjee and Jeevankumar 2007).

Three formations represent the Upper Vindhyan, viz. Kaimur, Rewa and Bhander in ascending order. All the formations have a lower shaley and/or calcareous part and an upper sandstone part (Fig. 1.3). The shaley Lower Kaimur is lenticular in geometry and confined in southeastern and eastern sectors of the Vindhyan outcrop area. With subordinate sandstone at the base, it generally fines upwards and represents deposition in a shallow marine condition (Table 1.1; Chakraborty and Bose 1990; Mandal et al. 2019). A basin-wide volcaniclastic deposit demarcates the base of the Upper Kaimur (Chakraborty et al. 1996). The basal part of the Upper Kaimur comprises a shallow shelf deposit that passes upwards gradually into fluvial and eolian deposits in a coarsening-upward succession (Chakraborty 1993). The Kaimur Formation passes upwards to the Rewa Formation with a basin-wide granular lag blanket in between (Bose et al. 2001). The lower part of the Rewa Formation consists of shale deposited in a storm-dominated shelf succession. Local kimberlite pipe-derived diamondiferous conglomerate occurs within the lower part of the Rewa Shale. The Rewa Shale passes upwards into the Rewa Sandstone, which is overall coarsening upwards. The sandstone shows an upward transition from shallow marine to fluvio-eolian deposits (Bose and Chakroborty 1994; Bose et al. 2001). The younger Bhander Formation in its lower part includes a laterally persistent carbonate deposit, the Bhander Limestone (Sarkar et al. 1996), bounded below and above, by interbedded mudstone–sandstone of the Ganurgarh Shale and the Lower Bhander Sandstone, respectively. Both the bounding members are red in colour and bear emergence features as well as salt pseudomorphs, probably indicating their coastal origin (Bose and Chaudhuri 1990; Chakraborty et al. 1998). The micritic, stromatolitic, intraclastic and oolitic Bhander Limestone represents deposition on a shelf (Sarkar et al. 1996, 1998, 2014a). The ~185-m-thick Sirbu Shale that overlies the Lower Bhander Sandstone, mostly indicates deposition on a storm-dominated offshore shelf (Sarkar et al. 2002b), barring its lower 7 m that is of lagoonal origin. The 110-m-thick progradational Upper Bhander Sandstone, at the top of the Vindhyan Supergroup, has a gradational contact with the underlying Sirbu Shale. In consequence, the basal part of the Upper Bhander Sandstone is a coastal deposit with occasional storm interventions, while its upper part belongs largely to a terrestrial environment, especially eolian (Bose et al. 1999; Sarkar et al. 2004).

Table 1.1 Depositional environment of the Vindhyan Supergroup exposed in the Son Valley area

	Formation	Member	Depositional environments		
Upper Vindhyan Group			Singh (1973)	Banerjee (1974)	Bose et al. (2001)
	Bhander	Upper Bhander Sandstone	Tidal flat—shoal complex	Tidal flat	Fluvio-eolian and marginal marine
		Sirbu Shale	Lagoon	Tidal flat (supratidal)	Lagoon to shelf
		Lower Bhander Sandstone		Tidal flat	Coastal playa
		Bhander Limestone	Carbonate tidal flat	Carbonate tidal flat	Shallow marine
		Ganurgarh Shale		Lagoon and tidal flat	Chenier
	Rewa	Rewa Sandstone	Shoal beach complex	Barrier beach—dune	Tidal to fluvio-eolian
		Rewa Shale	Lagoon	Lagoon	Shelf
	Kaimur	Upper Kaimur	Shoal complex	Barrier beach	Shelf in fluvio-eolian
		Lower Kaimur	Tidal flat	Barrier bar to tidal flat	Intertidal to shelf
Lower Vindhyan Group	Rohtas	Rohtas Limestone	Carbonate tidal flat (subtidal to intertidal)	Lagoon	Shelf
		Rampur Shale	Tidal flat (intertidal)	Carbonate tidal flat	Shelf
	Kheinjua	Chorhat Sandstone			Shallow marine
		Koldaha Shale	Lagoon	Lagoon to tidal flat	Dominantly shelf, deltaic fluvial
	Porcellanite		Lagoon	Lagoon	Shallow marine
	Kajrahat	Kajrahat Limestone	Carbonate tidal flat (subtidal to supratidal)	Lagoon (tidal flat towards top)	Subtidal to peritidal
		Arangi Shale	Lagoon	Shallow marginal lagoon	Shelf
	Deoland		High gradient coastal rivers	Mainland beach	Shallow shelf

1.4   Age of Vindhyan

The age of the Vindhyan Supergroup remains controversial because of apparent conflicting reports of radiometric dates and putative trace fossils (Venkatachala et al. 1996). The available age information till the last century brackets the age of the Vindhyan from 1200 to 550 Ma. However, all these radiometric dates rely on K–Ar, Rb–Sr and fission-track analyses, which may bear post-depositional or detrital signatures (Rasmussen et al. 2002). The report of triploblastic metazoans by Seilacher et al. (1998) from the Lower Vindhyan Semri Group has created tremendous interests in Vindhyan geology, particularly on chronostratigraphy and bio-sedimentology. Several new radiometric dates produced in the last 20 years now provide a much better age constraint, particularly for the Lower Vindhyan (Fig. 1.3).

The age of the Lower Vindhyan is firmly established based on SHRIMP U–Pb ages of magmatic zircons within the volcaniclastic units comprising the Porcellanite Formation of the Lower Vindhyan. Back-to-back publications of two broadly similar ages of 1628 ± 8 Ma (Rasmussen et al. 2002) and 1630.7 ± 0.8 Ma (Ray et al. 2002) redefine the age of the Porcellanite Formation as well as the Semri Group. Subsequently published ages of the Porcellanite Formation by U–Pb geochronology on magmatic zircons from rhyolite flows in the Porcellanite Formation have produced 1640 ± 4 Ma and 1647 ± 18 Ma (Bickford et al. 2017). These new radiometric dates firmly establish the age of the Porcellanite Formation between ~1640 and 1630 Ma. Kajrahat Limestone samples, about 150 m above the basement, has yielded a Pb–Pb isochron age of 1729 ± 110 Ma (Sarangi et al. 2004). These radiometric dates indicate the Vindhyan basin possibly opened at ca. 1800 Ma (Basu and Bickford 2015). An ash bed in the Rampur Shale of the Rohtas Formation yielded a U–Pb zircon age of 1599 ± 8 Ma (Rasmussen et al. 2002). The Pb–Pb isochron data for the same formation have produced 1599 ± 48 Ma (Sarangi et al. 2004). Another set of Pb–Pb isochron age for the Rohtas Limestone has produced 1514 ± 120 Ma (Chakrabarti et al. 2007). The equivalent of the same formation in the northern flank, the Tirohan Limestone has produced 1650 ± 89 Ma by Pb/Pb dating method (Bengtson et al. 2009). The Semri Group, therefore, brackets an age range between ~1800 and ~1500 Ma.

The age of the Upper Vindhyan is far more controversial, although much progress has been made in recent years towards resolving this issue (Fig. 1.3). The kimberlite pipes within Majhgawan and Hinota produced ages of 974 Ma to 1067 ± 31 Ma and 1170 ± 46 Ma by various techniques (Chalapathi Rao 2006). Gregory et al. (2006) reported a phlogopite Ar–Ar age of 1073.5 ± 13.7 Ma for the same kimberlite, indicating a Mesoproterozoic age for the oldest sediments of the Upper Vindhyan sequence. Glauconite samples in the Kaimur Group, providing a K–Ar age of 910 ± 39 Ma (Vinogradov et al. 1964), have been discarded (Chalapathi Rao 2006). Indirect evidence has supported a general view that the Rewa and Bhander groups belong to mid- to Late Neoproterozoic in age. These include a fission-track date of 710 ± 120 Ma for the Govindgarh Sandstone in the Upper Rewa Group (Srivastava and Rajagopalan 1988), as well as a comparison of Bhander Group carbonate Sr-isotope values to global patterns in the Neoproterozoic Era. Ray et al. (2003) suggested the ages of ~650 and 750 Ma for the Lakheri Limestone (Rajasthan) and the Bhander Limestone (Son Valley), respectively, based on Sr-isotope stratigraphy. However, similar isotope ratios show the age of Bhander limestone ~1.0 Ga (Gopalan et al. 2013). A mid- to Late Neoproterozoic age has also been bolstered by reports of enigmatic Ediacara-type fossils (De 2003, 2006) and large-diameter, burrow-like structures (Chakrabarti 1990) in the Bhander Group, although these findings have been strongly challenged by subsequent studies (Seilacher 2007; Banerjee et al. 2010; Kumar 2016). New evidence has emerged in recent years, suggesting a necessary revision of the age of the Upper Vindhyan sequence. Malone et al. (2008) found striking paleomagnetic similarities between the Bhander Group and the Majhgawan kimberlite, and suggested their similar age. They also found that the Upper Bhander Sandstone contained no detrital zircons younger than ~1020 Ma (see also Turner et al. 2014). Based on these lines of evidence, they considered the closure of the Vindhyan sedimentation by ~1000 Ma. Subsequently, Gopalan et al. (2013) reported Pb–Pb ages of 908 ± 72 Ma and 1073 ± 210 Ma for the Bhander and Lakheri limestones, respectively. Further, Tripathy and Singh (2015) reported a Re–Os age of 1210 ± 52 Ma for the Bijaigarh Shale in the Kaimur Group. The recent paleomagnetic and detrital zircon data, therefore, indicate that the Vindhyan Basin was closed around 900–1000 Ma (Malone et al. 2008; Gopalan et al. 2013; Basu and Bickford 2015). Therefore, on the basis of varied kinds of evidence, the most acceptable age of the Vindhyan Supergroup starts from ~1800 Ma to ~1000–900 Ma (Rasmussen et al. 2002; Ray et al. 2002, 2003; Ray 2006; Gopalan et al. 2013; Basu and Bickford 2015; Gilleaudeau et al. 2018).

1.5 Tectonic Setting of the Vindhyan Basin

1.5.1 Major Tectonic Elements

The Vindhyan Basin is bounded by Son–Narmada lineament in the southern margin, which constitutes a series of southerly dipping reverse faults (Fig. 1.1) (Kaila et al. 1985). This lineament cuts across the whole of Central India in NNE–SSW direction and has been periodically reactivated since the Precambrian (Choubey 1971; Kaila et al. 1985). The Son–Narmada lineament is a prominent linear feature in the Indian subcontinent with a total strike length of 1200 km and marks the boundary between the Aravalli–Bundelkhand Province and the Dharwar Province (Mazumder et al. 2000; Acharyya 2003). The deep seismic sounding studies suggest the existence of deep faults extending up to the Moho boundary along the Son–Narmada lineament (West 1962; Kaila et al. 1985, 1989; see also Acharyya 2003). The Son—Narmada South Fault (SNSF) generally delimits the northern boundary of the Gondwana Basins, whereas the Son–Narmada North Fault (SNNF) delineates the northern border of the Mahakoshal Mobile Belt and the southern boundary of the Vindhyan Basin (Acharyya 2003). The western margin of the Vindhyan basin is demarcated by a major tectonic lineament, consisting of a series of northeast–southwest trending, northwesterly dipping faults, called the Great Boundary Fault (GBF, Fig. 1.1; Tewari 1968; Naqvi and Rogers 1987; Narain 1987; Verma 1991; Srivastava and Sahay 2003) which runs for about 500 km. The lineament acted as a thrust boundary along which folded metasediments of the Paleoproterozoic Aravalli Supergroup, the Delhi Supergroup (1.9–1.4 Ga) and Bhilwara Supergroup (2.5–1.6 Ga), more popularly known as Delhi–Aravalli fold belt, lie against the Vindhyan Supergroup.

The crustal thickness of the Vindhyan Basin varies from 39.5 to 45 km (Kaila et al. 1989). Several wrench faults trending NE–SW and ENE–WSW directions divides the basin into several tectonic blocks (Das et al. 1999). The gravity anomaly and aeromagnetic studies on Vindhyan Basin reveal alternate mutually parallel basement highs and lows elongated in a general ENE–WSW to E–W directions (Verma and Banerjee 1992).

1.5.2 Basement Rocks

The Vindhyan Basin appears sickle-shaped on the Bundelkhand–Aravalli Province which stabilized prior to 2.5 Ga (Eriksson et al. 1999; Mazumder et al. 2000; Fig. 1.1). The Vindhyan Supergroup overlies on a variety of Precambrian basement rocks including Bundelkhand Granite, Mahakoshal Group, Bijawar Group, Gwalior Group, Banded Gneissic Complex (BGC) and Chhotanagpur Gneissic Complex (CGC). Bundelkhand Granite Complex separates the Vindhyan exposures of the Son Valley area from that of the Chambal Valley area (Fig. 1.1). The Bundelkhand Granite Complex is dominated by a K-rich granite emplacement within the Tonalite–Trondhjemite–Granodiorite Complex (Rogers 1986; Bandyopadhyay et al. 1995; Eriksson et al. 1999; Mazumder et al. 2000). The Mahakoshal Group (2.1–1.6 Ga) occurs on the southern and eastern margin of Son Valley (Das et al. 1990; Roy and Bandyopadhyay 1990; Nair et al. 1995). Bijawar Group (2.1 Ga) remains confined on the northern and northeastern margin of the Son Valley (Das et al. 1990; Roy and Bandyopadhyay 1990). Banded Gneissic Complex (BGC) and Gwalior Group form the basement of the Vindhyan Supergroup in the Rajasthan sector (Heron 1953). Chhotanagpur Gneissic Complex (CGC) consisting of gneisses, granites and granodiorites with enclaves of tonalitic gneisses and ultramafics forms the basement of the Vindhyan Basin in most parts of southeastern Son Valley area (Singh et al. 2001).

1.5.3 Tectono-Sedimentation Model

Different ideas have been proposed about the tectonic setting of the Vindhyan Basin (Table 1.2). The southern part of the basin is locally concealed under the Deccan lava that erupted during the end of Cretaceous. The Vindhyan succession overlies the early Proterozoic metasediments of Bijawar and Mahakoshal Groups and underlies

Table 1.2 Tectonic model proposed by previous workers for the Vindhyan Basin

Tectonic models	Authors
Foreland basin	Auden (1933)
Foreland basin	Chakraborty and Bhattacharyya (1996)
Foreland basin	Raza and Casshyap (1996)
Foreland basin	Chakrabarti et al. (2007)
Strike–slip basin	Crawford (1978)
Syncline	Chanda and Bhattacharyya (1982)
Syncline	Chaudhuri and Chanda (1991)
Syncline	Prakash and Dalela (1982)
Rift basin	Choubey (1971)
Rift basin	Naqvi and Rogers (1987)
Rift basin	Kaila et al. (1989)
Rift basin	Verma and Banerjee (1992)
Rift basin	Ram et al. (1996)
Intracontinental platform basin	Valdiya (1982)
Rift to sag Transition	Bose et al. (2001)

the Gondwanas in central India. Since the Aravalli, Delhi and Satpura orogenic belts border it, some workers considered the Vindhyan Basin as a peripheral foreland basin related to the southerly dipping subduction before the collision of the Bhandara and Bundelkhand cratons (Auden 1933; Chakraborty and Bhattacharyya 1996; Raza and Casshyap 1996). Sedimentation in a foreland basin verging northwards (Chakraborty and Bhattacharyya 1996) or southwards (Chakrabarti et al. 2007) was suggested. Some workers envisaged the Vindhyan Basin as a strike–slip fault basin (Crawford and Compston 1970; Crawford 1978). Chakrabarti et al. (2007) supported the idea of a foreland based on Nd isotope study of clastic and non-clastic sedimentary deposits within the Vindhyan succession. A different view postulated an intracratonic rift origin for the Vindhyan Basin (Verma and Banerjee 1992; Ram et al. 1996). However, the overall fine grain size and high textural and mineralogical maturity of sandstones defy rapid sedimentation from supracrustal source and do not comply with these suggestions. Bose et al. (2001) correlated the sedimentary and geophysical attributes to an intracratonic rift to sag transition.

A broad consensus is that the Vindhyan sediments were deposited within an E–W elongated and westward opening intracratonic basin (Chanda and Bhattacharyya 1982; Chaudhuri and Chanda 1991; Sarkar et al. 2004; Chakraborty et al. 2012). Bose et al. (2001, 2015) recorded initial rift (Lower Vindhyan) to sag stage (Upper Vindhyan) of evolution within the Vindhyan Basin. Extensive studies on multiple fronts later reveal intracratonic north–south rifting with a dextral shear at the initial stage (Bose et al. 1997, 2001) and sag at a subsequent stage (Sarkar et al. 2002b). Consequently, the east–west-elongated main Vindhyan Basin had initially been divided into several sub-basins by several NW–SE-oriented ridges (Bose et al. 1997), but during the Upper Vindhyan sag stage, this segmentation was largely removed (Bose et al. 2001). General confinement of syn-depositional faults within the Semri succession corroborates the change in tectonic setting from the Lower to the Upper Vindhyan time (Ram et al. 1996). Further, volcaniclastics are important constituents of the Semri Group, but they are insignificant components within the Upper Vindhyan (Chakraborty et al. 1996).

1.6 Biotic Records within the Vindhyan Supergroup

The Vindhyan sediments have consistently produced evidence of life since the beginning of the twentieth century. Gently metamorphosed sandstones and carbonates of the Vindhyan Supergroup yield a wide variety of paleobiological reports. The fossil reports of the Vindhyan Supergroup include microfossils, organo-sedimentary structures, carbonaceous fossils, Ediacaran fossils, trace and body fossils of metazoans (Tables 1.3, 1.4, and 1.5). Some of these fossils broadly constrain the age of the Vindhyan Supergroup. At the same time, many fossil reports provide conflicting ages. Apart from biostratigraphic significance, the fossil reports illustrate the evolution of microbial and metazoan life in Earth's history. Paleobiological remains of the Vindhyan Supergroup belong to separate categories, viz. stromatolites, megascopic carbonaceous remains, microfossils, small shelly fossil, pseudo-Ediacaran fossils and microbially induced sedimentary structures (MISS).

1.6.1 Stromatolite

Stromatolites or microbialites are organo-sedimentary deposits that indicate the interaction between benthic microbial communities and detrital or chemical sediments. Various workers used stromatolites and other organo-sedimentary structures for biostratigraphy since many of the structures are time- and space-constrained (Raaben 1969, 2005; Kumar 1978, 1980; Raha and Sastri 1982). Distinctive *Kussiella, Conophyton, Colonnella* assemblages occur within the Vindhyan Supergroup (Table 1.3). While *Conophyton, Colonnella* and *Kussiella* characterize the Semri Group, *Baicalia* and *Tungussia* occur in the Bhander Group. *Conophyton* was conspicuously absent in the latter. Based on stromatolitic assemblages, a few workers proposed Early to Middle Riphean age (1600–1400 Ma and 1400–100 Ma) for the Semri Group and Middle Riphean to Upper Riphean (1000–650 Ma) for the Upper Vindhyan Group (Kumar 1984; Kumar et al. 2005). While *Conophyton* strictly occurs in the Precambrian, *Cryptozoon* can also be found in modern intertidal seas (Logan 1961). Subsequently, many workers found considerable environmental control on stroma-

Table 1.3 Stromatolites reported from the Vindhyan Supergroup and its age implications

Member/Formation	Author	Stromatolite	Proposed age
Upper Bhander Sandstone			
Sirbu Shale	Prasad and Ramaswamy (1980) Prasad (1984)	*Collenia baicalica* *Collenia columnari* *Collenia buriatica* *Collenia baicalica*	Middle to Upper Riphean Middle to Upper Riphean
Bhander Limestone	Rao et al. (1977) Kumar (1976a) Sarkar (1974) Kumar (1982)	*Collenia symmetrica* *Baicalia baicalica* *Maiharia maiharensis* *Collenia undusa* *Tungussia, Boxonia*	Upper Riphean (900–600 Ma) Upper Riphean Middle Riphean Upper Riphean
Rohtas Limestone	Sharma (1996) Singh and Banerjee (1980) Valdiya (1969)	*Collenia kusiensis* *Collenia columnaris* *Conophyton cylindricus* *Collenia kusiensis* *Collenia baicalica*	Middle Riphean Middle Riphean Middle Riphean Middle Riphean
Koldaha Shale	Kumar (1976b)	*Collenia cylindrica* *Colonella columnaris* *Collenia clappii*	Lower Riphean Lower Riphean Lower Riphean
Kajrahat Limestone	Prasad (1980) Prasad (1984) Kumar (1982) Kumar and Gupta (2002)	*Conophyton cylindricus* *Collenia kusiensis* *Cryptozoon occidentale* *Colonnella Sp.* *Collenia frequence* *Conophyton inclinatum* *Kussiella kusiensis* *Kussiella, Colonnella*	Lower Riphean Lower Riphean Lower Riphean Lower Riphean Lower Riphean Lower Riphean

tolite morphology and questioned the biostratigraphic application of stromatolites (Davaud et al. 1994; see also Altermann 2002, 2004).

1.6.2 *Carbonaceous Megafossils*

Various kinds of carbonaceous fossils were reported from the Vindhyan Supergroup (Table 1.4). Most of these fossils were found within the Rohtas Formation and its equivalent in western Vindhyan, the Suket Shale. Jones (1909) first reported small carbonized spherical discs from the Suket Shale near Rampura. This discovery received the attention of many workers, and the debate regarding the exact affinity of these fossils continued for many decades. Chapmann (1935) assigned the specimens into two new genera and four new species belonging to *Protobolella jonesi, Fermoria*

Table 1.4 Carbonaceous fossil reports from the Vindhyan Supergroup

Carbonaceous megafossils	Stratigraphic horizon	Author(s)	Remarks[a]
Carbonaceous discs described as *Obolella*/*Chuaria*/Operculam of *Hyolithella*	Suket Shale~ Rampur Shale	Jones (1909)	Revised as *Chuaria circularis*
Spiral impression described as impression of coiled worm	Rohtas Limestone	Beer (1919)	Revised as *Grypania spiralis*
Fermoria minima, Fermoria granulosa, Fermoria capsella, Protobolella jonesi described primitive brachiopod or eurypterids	Suket Shale~Rampur Shale	Chapmann (1935)	Revised as *Chuaria circularis*
Vindhyanella jonesi, Fermoria minima	Suket Shale	Sahni (1936)	*Revised as Chuaria circularis*
Carbonaceous discs and algal dust	Rohtas Limestone	Misra and Bhatnagar (1950)	*Revised as Amjhorea rohtasae*
Circular to semi ovoid carbonaceous discs	Suket Shale	Misra and Dube (1952)	*Chuaria circularis*
Circular form *Fermoria* sp. broadly ovoid form	Suket Shale	Misra (1957)	*Chuaria circularis*
Fermoria-like structures	Sirbu Shale	Misra and Awasthi (1962)	Uncertain
Katnia singhi classified under Annelida	Rohtas Limestone	Tandon and Kumar (1977)	Similar to *Grypania*
'disc-like remains type 1–6'	Suket Shale	Maithy and Shukla (1977)	Considered as *Chuaria circularis*
Vavosphaeridium reticulatum, Vavosphaeridium vindhyanesi, Kildinella suketensis, Tasmanites vindhyanesis	Suket Shale	Maithy and Shukla (1977)	Revised as *Chuaria circularis*
Chuaria circularis, Tawuia suketensis, Vindhyania jonesii	Suket Shale	Mathur (1982)	*Vindhyania jonesii is similar to Krishnania acuminate*
Chuaria circularis, Chuaria fermorei Tawuia suketensis, Tawuia rampuraensis	Suket Shale	Mathur (1982)	Revised as *Chuaria circularis* Revised as *Tawuia dalensis*
Chuaria minima	Suket Shale	Maithy and Shukla (1984)	
Chuaria minima Tawuia dalensis	Rohtas Limestone	Maithy and Babu (1988)	

(continued)

Table 1.4 (continued)

Carbonaceous megafossils	Stratigraphic horizon	Author(s)	Remarks[a]
Tyrasotaenia sp.	Suket Shale	Shukla and Sharma (1990)	
Krishnania acuminata, Krishnania multistriata	Rohtas Limestone	Maithy (1991)	*Krishnania* possibly represents oldest benthic algal forms
Chuaria circularis, Grypania sp.	Rohtas Limestone	Kumar (1995)	
Chuaria circularis, Tawuia dalensis	Bhander Limestone and Sirbu Shale	Kumar and Srivastava (1997)	
Chuaria circularis, Tawuia dalensis	Rewa Shale	Rai et al. (1997)	
Chuaria circularis, Chuaria gigantia, Chuaria melanocentrics, Grypania spiralis, Grapania sp., *Phyllonia bistaria*	Rohtas Limestone	Rai and Gautam (1998)	*Chuaria gigantia* and *Chuaria melanocentrics* are the synonymy of *Chuaria circularis*
Chuaria circularis, Tawuia dalensis, Chambalia minor, Beltina danai, Chuaria vindhyanensis	Rohtas Limestone	Kumar (2001)	
Chuaria circularis, tawuia dalensis	Rewa Shale	Srivastava (2004)	

[a]Remarks regarding affinity of the carbonaceous fossils after Venkatachala et al. (1996), Banerjee (1997), Kumar (2016), Sharma et al. (2012)

minima, F. granulosa and *F. capsella*. Sahni (1936) placed all of them in the synonymy of *F. minima* but created another name *Vindhyanella* for one of the specimens figured as *Protobolella jonesi* by Chapmann (1935, pl. 2, Fig. 1). Misra (1957), however, considered that *Fermoria* were chlorite aggregates in schist and others were hematite spots in sandstone. Maithy and Shukla (1977) suggested that the disc-like bodies represent either acritarchs (eukaryotic algae of unknown affinity or algal colonies). Maithy and Shukla (1984) considered these forms as *Chuaria circularis*. They compared the specimens with cryptarch genus *Orygmatosphaeridium*. Maithy and Babu (1988) reported a tubular variety of carbonaceous fossils and considered them *Tawuia*. Kumar (2001) suggested that the carbonaceous fossils represent different parts of a multicellular Chlorophycean/Xanthophycean plant. *Chuaria circularis* represents a compressed cyst-like spherical body which was attached with a siphonaceous/filamentous thallus (*Tawuia*). Although the *Chuaria* fossils were considered morphologically similar to *Chuaria circularis*, the biological affinity is still unknown. Currently, there are at least three possible interpretations of *Chuaria circularis*.

Table 1.5 Microfossils reported from Vindhyan Supergroup

Formation	Member	Author	Evidence	Proposed age
Bhander	Upper Bhander Sandstone			
	Sirbu Shale	Maithy and Mandal (1983)	Archaeorestis sp.	Late Proterozoic
		Maithy and Meena (1989)	Spaerophycus pervum	Late Proterozoic
			Myxoccoides psilata	
			Protosphaeridium desum	
	Lower Bhander Sandstone			
	Bhander Limestone	Maithy and Gupta (1983)	Biocatenoides sphaerula	Late Proterozoic
			Sphaerophycus parvum	
	Ganurgarh Shale			
Rewa	Rewa Sandstone			
	Rewa Shale	Maithy and Mandal (1983)	Myxoccoides psilata	Late Proterozoic
			Nanococcus vulgaris	
Kaimur				
Rohtas	Rohtas Limestone	Maithy and Shukla (1977)	Myxoccoides globosa	Late Riphean
			Palaeonacystis suketensis	
		Maithy and Babu (1988)	Misraea psilata	1200–940 Ma
		Nautiyal (1986)	Leisphaeridia densum	1000–940 Ma
Kheinjua	Chorhat Sandstone		Eomycetopsis septata	
	Koldaha shale(Fawn Limestone)	Kumar and Srivastava (1995)	Sphaerophycus parvum	1200 Ma
			Bontophysalis belcherensis	
			Oscillatorioppsis breviconvera	
		McMenamin et al. (1983)	Myxoccoides minor	1200 Ma
			Tetraphycus congregatus	

(continued)

Table 1.5 (continued)

Formation	Member	Author	Evidence	Proposed age
		Nautiyal (1983)	Kildinella aff.	1200–900 Ma
Porcellanite				
Kajrahat		Nautiyal (1983)	Palaeoanacystis suketensis	1200–900 Ma
			Eomycetopsis filiformis	

(1) Several authors classify *C. circularis* in Acritarcha (Evitt 1963), compare it with *Leiosphaeridia* and interpret it as a eukaryotic algal cyst (e.g. Ford and Breed 1973).

(2) The coexistence of the transitional forms between *C. circularis* and *Tawuia dalensis* indicates their close affinities, representing vegetative stages of a (multicellular) eukaryotic alga (Duan 1982; Kumar 2001; Dutta et al. 2006; Sharma et al. 2009, 2012).

(3) *C. circularis* and *T. dalensis* are colonial cyanobacteria, similar to *Nostoc* ball (Sun 1987). Steiner (1997) reached a similar conclusion, although he admitted that some *C. circularis* populations may be eukaryotic.

Tandon and Kumar (1977) described two carbonaceous fossils *Katnia* and *Vindhyavasini* from Katni area within the Rohtas Limestone, which were considered by them as annelids and arthropods, respectively. Kumar (1995) reinterpreted *Katnia* as *Grypania* (see also Kumar 2001), which was first reported by Beer (1919) from the study area near Amjhore. *Grypania* is regarded as probable eukaryotic alga because of its complexity, structural rigidity and large size (Han and Runnegar 1992). Such fossils occur in rocks as old as 2.1 Ga (Han and Runnegar 1992). *Chuaria* was considered as index fossils of the Neoproterozoic (Ford and Breed 1973; Sun 1987). However, it was reported from the Paleoproterozoic (Hofmann and Chen 1981; Du and Tian 1985; Zhu et al. 2000). Steiner (1997) and Amard (1997) reported *Chuaria* fossils from Early Cambrian. Therefore, *Chuaria circularis* possibly existed since Paleoproterozoic and continued up to Early Cambrian and appeared to be an insignificant biostratigraphic tool for the global correlation of Proterozoic sedimentary sequences.

1.6.3 Microfossils

A large number of microfossils were reported from both argillaceous rocks and carbonates of the Vindhyan Supergroup (Table 1.5; also see Venkatachala et al. 1996). Most of these microfossils belonged to cyanobacteria, and only a few of them belonged to eukaryotic acritarch. In general, the microfossil record suggested Mesoproterozoic age for the Semri Group and Neoproterozoic age for the Upper Vindhyan Group. Sharma and Sergeev (2004) reported typical Mesoproterozoic microbiotas from the Kheinjua Formation. Some of the microfossil data, however, provided anomalous age (e.g. Srivastava 1971). Because of minute size, partial preservation and simple morphologies, such fossils could be confused with non-biologic microstructures (Schopf 1975, 2004; Cloud 1976; Schopf and Walter 1983; Brasier et al. 2002).

1.6.4   Ediacara-like Fossils

The Vindhyan Supergroup has yielded probable Ediacaran fossils such as *Ediacaria flindersi, Cyclomedusa davidi, Medusinites, Medusinites asteroides, Dickinsonia* and *Beltanelliformis brunsae.* Kathal et al. (2000) reported an Ediacaran genus from the Palkawan Shale (equivalent to Porcellanite Formation) of Semri Group, near Sagar, which was compared with *Spriggina floundersi.* Kathal et al. (2000) suggested a re-evaluation of the age of the Vindhyans. However, Kumar (2001) contested both these reports and attributed the structures to weathering features. Williams and Schmidt (2003) and De (2003) reported problematic Ediacaran fossils from the Semri Group and the Bhander Formation. Williams and Schmidt (2003) compared the fossils impression with *Rugoconites*, a medusoid of uncertain affinity, whereas De (2003) attributed these fossils to *Ediacaria* sp. and *Hiemalora* sp. The biogenicity of both fossil reports is questionable because of the recent radiometric dating and because of their resemblance with microbial mat-related structures (Seilacher 1997; Banerjee et al. 2010).

1.6.5   Problematic Body and Trace Fossils of Metazoans

Many workers reported trace fossils from the Vindhyan sedimentary succession. Most common of these structures include meandering features found in the trough of wave ripples, considered as *Muniaichnus* (Kumar 1978) and *Cochlichnus Hitchcock* (Kulkarni and Borkar 1996a). Sarkar et al. (2004) attributed such features as syneresis cracks on microbially colonized sandy bed surfaces (see also Hofmann 1971; Seilacher et al. 1998). The fossil reports of Misra and Awasthi (1962), Mathur (1983), Singh and Sinha (2001) were compared with cracks (Kumar 2001). Chakrabarti (2001) considered the meandering and sinuous structures of the Vindhyan as dubiofossils. Sandy sediments of Proterozoic sea floors, covered by the moneran carpet could turn cohesive enough to form cracks (Banerjee 1997; Seilacher et al. 1998; Schieber 1998, 2004; Sarkar et al. 2004). Other reported fossils possibly owe their origin to sand volcanoes and other water escape features (e.g. burrows of Kulkarni and Borkar 1996b), gas escape structures and impressions of algae and microbial mat on the sandy substrate (Seilacher 1997).

Azmi (1998) reported small shelly fossils from the phosphoritic stromatolitic dolomite in the basal part of the Vindhyan Supergroup from Chitrakut area and claimed Cambrian age. Bengtson et al. (2009) reinterpreted the fossils as filamentous and coccoid cyanobacteria and filamentous eukaryotic algae. Bengtson et al. (2017) further considered the fossils as the earliest known crown group of eukaryotes based on preserved cellular and subcellular diverse forms showing affinity to rhodophytes and renamed these fossils as *Rafatazmia* and *Ramathallus*.

Discovery of traces of possible metazoans from the Kheinjua Formation by Sarkar et al. (1996) and Seilacher et al. (1998) created a sensation in Precambrian paleobiology. Subsequently, metazoan fossils were discovered from contemporary Proterozoic rocks (Rasmussen et al. 2002, 2004). All these findings questioned the 'Cambrian explosion' theory of metazoan divergence. The claim made by Seilacher et al. (1998) was consistent with the results of molecular geneticists (Wray et al. 1996; Ayala et al. 1998; Wang et al. 1999). DNA sequence analysis and molecular clocks do not support Cambrian explosion (Blair and Hedges 2005). It considers basal animal phyla (Porifera, Cnidaria) possibly diverged between 1200 and 1500 Ma (Blair and Hedges 2005). The biogenicity of the Chorhat fossils and early metazoan evolution has remained a contentious issue (Conway Morris 2000, 2003; Knoll 2003; Fedonkin 2003; Peterson et al. 2005). Meanwhile, Seilacher (2007) retracted from his original interpretation because of the recent revision of age of the Chorhat sandstone to 1.6 Ga.

1.6.6 Microbially Related Structures (MRS)

The Vindhyan Supergroup is well known globally for superb preservation of delicate microbial structures on siliciclastics (Sarkar et al. 2004, 2005, 2006, 2014b, 2016; Banerjee and Jeevankumar 2005; Banerjee et al. 2006a, 2010, 2014; Sarkar and Banerjee, 2007; Schieber et al. 2007; Eriksson et al. 2010). Carbonaceous shale (total organic carbon content exceeding 1.5%) associated with the condensed zones exhibits wavy, crinkly, carbonaceous laminae, pyritic laminae, pseudo-cross-strata, rolled-up and folded carbonaceous laminae in Rampur shale, Sirbu Shale and in Kajrahat Formation, suggesting microbial mat growth in mid- to outer shelf depositional conditions (Banerjee et al. 2006a; Sur et al. 2006). Banerjee and Jeevankumar (2005) and Sarkar et al. (2004, 2005, 2006) recorded several microbial structures within shales and sandstone beds across the Vindhyan succession. The best presentation of MRS has been found in shallow subtidal to supratidal environment within the Chorhat Sandstone and the Upper Bhander Sandstone (Sarkar et al. 2006, 2016; Bose et al. 2007; Schieber et al. 2007). Detailed discussion regarding microbially related structures is provided in Chap. 5.

1.7 Igneous Intrusives

The Vindhyan succession is intruded by a few mafic and felsic intrusive bodies. A large number dolerite dikes and sills occur within the Semri Group (Auden 1933; Ahmad 1971; Srivastava and Iqbaluddin 1981), the Rewa Formation (Soni et al. 1987 and references therein) and the Bhander Formation (Soni et al. 1987). Several diamondiferous ultramafic pipes intrude the Kaimur Formation in Majhgawan and Hinota in the Panna area (Mathur and Singh 1971; Kailasam 1979; Paul 1991; Ravi Shanker et al. 2001; Chalapathi Rao 2005, 2006). The Majhgawan pipe occurs on the western limit of the Panna diamond belt (80 × 50 km) and is localized in a NE–SW to ENE–WSW trending crestal zone of the upwarped eastern margin of the Bundelkhand craton (Halder and Ghosh 1978). The Hinota pipe is a circular intrusion with a shallow crater of up to 80 m (Chalapathi Rao 2006). The Mesoproterozoic diamondiferous ultramafic pipes at Majhgawan and Hinota, which intrude the Kaimur Group of Vindhyan rocks, combine the petrological, geochemical and isotope characteristics of kimberlite, orangeite (Group II kimberlite) and lamproite, and hence are characterized as belonging to 'transitional kimberlite orangeite–lamproite' rock type. While the Hinota pipe produced 1170 ± 46 Ma age by K/Ar technique (Paul et al. 1975), the Majhgawan kimberlite yielded 974 Ma to 1067 ± 31 Ma by various techniques (Chalapathi Rao 2006).

1.8 Volcanism

In Rajasthan area, Semri sedimentation commenced with contemporaneous volcanic activity as indicated by andesitic tuff, pyroclastics and breccias, formally known as Khairmalia Pyroclastics (1250 Ma; Crawford and Compston 1970; Prasad and Verma 1991; Raza et al. 2001). The Khairmalia volcanics has no equivalent in the Son Valley (Prasad 1984). The Semri Group in the Son Valley is distinctive from Upper Vindhyans by having huge piles of volcanic materials (mainly Porcellanite Formation). Porcellanite Formation is mostly pyroclastic in nature (Auden 1933; Ghosh 1971; Chakraborty et al. 1996; Banerjee 1997; Rasmussen et al. 2002). Volcaniclastics of the Porcellanite Formation occur in the form of surges, tuffs and epiclastics (Banerjee 1997). Mishra et al. (2018) considered Plinian-type eruptions from isolated events along a 300-km crustal fracture in the Son Valley region. Different opinions exist regarding the depositional setting of the Porcellanite Formation. Srivastava (1977) recognized subaerial deposition, while Banerjee (1997) considered frequent transitions from subaqueous to subaerial conditions. Felsic volcanic materials occur at specific levels within Kaimur and Rewa Formations (Chakraborty et al. 1996).

References

Acharyya SK (2003) The nature of Mesoproterozoic Central Indian Tectonic Zone with exhumed and reworked older granulites. Gond Res 6:197–214

Ahmad F (1958) Paleogeography of central India in the Vindhyan Period. Rec Geol Surv India 87:513–548

Ahmad F (1971) Geology of the Vindhyan System in the eastern part of the Son Valley in Mirzapur District, U.P. Rec Geol Surv India 96:1–41

Akhtar K (1996) Facies, sedimentation processes and environments in the Proterozoic Vindhyan Basin. In: Bhattacharyya A (ed) Recent Advances in Vindhyan Geology, vol 36. Mem Geol Soc India, pp 127–136

Altermann W (2002) The evolution of life and its impact on sedimentation. In: Altermann W, Corcoran PL (eds) Precambrian Sedimentary Environments: a Modern Approach to Ancient Depositional Systems, vol 33. Spec Publ Int Ass Sed, Blackwell, Oxford, pp 15–32

Altermann W (2004) Precambrian stromatolites: problems in definition, classification, morphology and stratigraphy. In: Eriksson PG, Altermann W, Nelson DR, Muller WU, Catuneanu O (eds) The precambrian Earth: Tempos and Events. Elsevier, Amsterdam, pp 639–670

Amard B (1997) Chuaria pendjariensis n. sp. Acritarche du bassin des Volta, Benin et Burkina–Faso, Africa de l'Ouest: un taxon nouveau du Cambrien inférieur. C R Acad Sci Paris T324 (série Iia), 477–483

Auden JB (1933) Vindhyan sedimentation in the Son Valley, Mirzapur District. Mem Geol Surv India 62:140–250

Ayala FJ, Rzhestky A, Ayala FJ (1998) Origin of the metazoan phyla: molecular clocks confirm paleontological estimates. Proc Nat Acad Sci USA 95:606–611

Azmi RJ (1998) Discovery of lower Cambrian small shelly fossils and brachiopods from the Lower Vindhyan of Son Valley, Central India. J Geol Soc India 52:381–389

Bandyopadhyay B, Roy A, Huin AK (1995) Structure and tectonics of a part of central Indian Shield. In: Sinha Roy S, Gupta KR (eds) Continental crust of the Northwestern and Central India, vol 3. Mem Geol Soc India, pp 433–467

Banerjee I (1964) On some broader aspects of Vindhyan sedimentation. Report 22nd International Geological Congress, Delhi, vol 15, pp 189–204

Banerjee I (1974) Barrier coastline sedimentation model and the Vindhyan example. Quart J Geol Min Met Soc India 46:101–127

Banerjee S (1997) Facets of Mesoproterozoic Semri sedimentation, Son valley, MP. Unpubl PhD Thesis, Jadavpur University, Kolkata, India

Banerjee S (2000) Climatic versus tectonic control on storm cyclicity in Mesoproterozoic Koldaha Shale, central India. Gond Res 3:521–528

Banerjee S (2010) Distinction between marine and continental facies in Precambrian sedimentary succession: Palaeoproterozoic Deoland Formation, Vindhyan Supergroup, central India. Gond Geo Mag 25:239–250

Banerjee S, Jeevankumar S (2003) Facies motif and paleogeography of Kheinjua Formation, Vindhyan Supergroup, Eastern Son valley. Gond Geol Mag (Spec Pub) 7:363–370

Banerjee S, Jeevankumar S (2005) Microbially originated wrinkle structures on sandstone and their stratigraphic context: Palaeoproterozoic Koldaha Shale, central India. Sed Geol 176:211–224

Banerjee S, Jeevankumar S (2007) Facies and depositional sequence of the Mesoproterozoic Rohtas Limestone: Eastern Son valley, India. J Asian Earth Sci 30:82–92

Banerjee S, Sarkar S, Bhattacharyya SK (2005) Facies, dissolution seams and stable isotope characteristics of the Rohtas Limestone (Vindhyan Supergroup) in the Son valley area, central India. J Earth Sys Sci 114:87–96

Banerjee S, Dutta S, Paikaray S, Mann U (2006a) Stratigraphy, sedimentology and bulk organic geochemistry of black shales from the Proterozoic Vindhyan Supergroup (central India). J Earth Sys Sci 115:37–48

Banerjee S, Jeevankumar S, Sanyal P, Bhattacharyya SK (2006b) Stable isotope ratios and nodular limestone of the Proterozoic Rohtas Limestone: Vindhyan Basin, India. Carbonates Evaporites 21:133–143

Banerjee S, Bhattacharya SK, Sarkar S (2006c) Carbon and oxygen isotope compositions of the carbonate facies in the Vindhyan Supergroup, central India. J Earth Sys Sci 115:113–134

Banerjee S, Bhattacharya SK, Sarkar S (2007) Carbon and oxygen isotopic variations in peritidal stromatolite cycles, Paleoproterozoic Kajrahat Limestone, Vindhyan Basin of central India. J Asian Earth Sci 29:823–831

Banerjee S, Jeevankumar S, Eriksson PG (2008) Mg–rich ferric illite in marine transgressive and highstand systems tracts: examples from the Paleoproterozoic Semri Group, central India. Precam Res 162:212–226

Banerjee S, Sarkar S, Eriksson PG, Samanta P (2010) Microbially related structures in siliciclastic sediment resembling Ediacaran fossils: examples from India, ancient and modern. In: Seckbach J, Oren A (eds) Microbial Mats: Modern and Ancient Microorganisms in Stratified System. Springer, Berlin, pp 111–129

Banerjee S, Sarkar S, Eriksson PG (2014) Palaeoenvironmental and biostratigraphic implications of microbial mat–related structures: examples from modern Gulf of Cambay and Precambrian Vindhyan Basin. J Paleogeography 3:127–144

Basu A, Bickford ME (2015) An alternate perspective on the opening and closing of the intracratonic Purana basins in peninsular India. J Geol Soc India 85:5–25

Basumallick S, Sarkar BC, Banerjee S (1996) Tidal cyclicity in Lower Bhander Sandstone, Maihar, Madhya Pradesh. J Geol Soc India 47:189–194

Beer EJ (1919) Note on spiral impression on Lower Vindhyan limestone. Rec Geol Surv India 50:139

Bengtson S, Belivanova V, Rasmussen B, Whitehouse M (2009) The controversial "Cambrian" fossils of the Vindhyan are real but more than a billion years older. Proc Nat Acad of Sci USA 106:7729–7734

Bengtson S, Sallstedt T, Belivanova V, Whitehouse M (2017) Three-dimensional preservation of cellular and subcellular structures suggests 1.6 billion-year-old crown-group red algae. PLoS Biol, 1–38. https://doi.org/10.1371/journal.pbio.2000735

Bhattacharyya A (1996) Proterozoic Kaimur Group, Son Valley: fluvio–marine or fluvio–lacustrine? J Geol Soc India 47:313–324

Bhattacharyya A, Morad S (1993) Proterozoic braided ephemeral fluvial deposits: an example from the Dhandraul Sandstone Formation of the Kaimur Group, Son valley, central India. Sed Geol 84:101–114

Bickford ME, Mishra M, Muelle PA, Kamenov GD, Schieber J, Basu A (2017) U–Pb age and Hf isotope compositions of magmatic zircons from a rhyolite flow in the Porcellanite Formation in the Vindhyan Supergroup, Son Valley (India): implications for its tectonic significance. J Geol 125:367–379

Blair JE, Hedges SB (2005) Molecular clocks do not support the Cambrian explosion. Mol Biol Evol 22:387–390

Bose PK, Chaudhuri AK (1990) Tide versus storm in epeiric coastal deposition: two Proterozoic sequences, India. Geol Jour 25:81–100

Bose PK, Chakroborty PP (1994) Marine to Fluvial transition: Proterozoic Upper Rewa sandstone, Maihar, India. Sed Geol 89:285–302

Bose PK, Banerjee S, Sarkar S (1997) Slope–controlled seismic deformation and tectonic framework of deposition of Koldaha Shale, India. Tectonophys 269:151–169

Bose PK, Chakraborty S, Sarkar S (1999) Recognition of ancient eolian longitudinal dune: a case study in Upper Bhander Sandstone, Son valley, India. J Sed Res 69:86–95

Bose PK, Sarkar S, Chakraborty S, Banerjee S (2001) Overview of the Meso to Neoproterozoic evolution of the Vindhyan Basin, central India. Sed Geol 141:395–419

Bose PK, Sarkar S, Banerjee S, Chakraborty S (2007) Mat–related features from the Vindhyan Supergroup in central India. In: Schieber J, Bose PK, Eriksson PG, Banerjee S, Sarkar S,

Catuneanu O, Altermann W (eds) An Atlas of Microbial Mat Features Preserved Within the Clastic Rock Record. Elsevier, pp 181–188

Bose PK, Sarkar S, Das NG, Banerjee S, Mandal A, Chakraborty N (2015) Proterozoic Vindhyan Basin: configuration and evolution. Mem Geol Soc London 43:85–102

Brasier MD, Green OR, Jephcoat AP, Kleppe AK, Van Kranendok MJ, Lindsay JF, Steel A, Grassineau NV (2002) Questioning the evidence of earth's oldest fossils. Nature 416:76–81

Chakrabarti A (1990) Traces and dubiotraces: examples from the so–called Late Proterozoic silici-clastic rocks of the Vindhyan Supergroup around Maihar, India. Precam Res 47:41–153

Chakrabarti A (2001) Are meandering structures found in Proterozoic rocks of different ages of the Vindhyan Supergroup of central India biogenic? A scrutiny. Ichnos 8:131–139

Chakrabarti R, Basu AR, Chakrabarti A (2007) Trace element and Nd–isotopic evidence for sediment sources in the mid–Proterozoic Vindhyan Basin, central India. Precam Res 159:260–274

Chakraborty C (1993) Morphology, internal structure and mechanics of small longitudinal (seif) dunes in an aeolian horizon of Proterozoic Dhandraul Quartzite, India. Sedimentol 40:79–85

Chakraborty C (1995) Gutter casts from the Proterozoic Bijaigarh Shale Formation, India: their implications for storm–induced circulation in shelf settings. Geol J 30:69–78

Chakraborty C (2001) Lagoon–tidal flat sedimentation in an epeiric sea: Proterozoic Bhander Group, Son Valley, India. Geol J 36:125–141

Chakraborty C, Bose PK (1990) Internal structures of sand waves in a tide–storm interactive system: Proterozoic Lower Quartzite Formation, India. Sed Geol 67:133–142

Chakraborty C, Bose PK (1992) Rhythmic shelf storm beds: Proterozoic Kaimur Formation, India. Sed Geol 77:259–268

Chakraborty C, Bhattacharyya A (1996) Fan delta sedimentation in a foreland moat: Deoland Formation. Geol Soc India Mem 36:27–48

Chakraborty PP (2004) Facies architecture and sequence development in a Neoproterozoic carbonate ramp: Lakheri Limestone Member, Vindhyan Supergroup, central India. Precam Res 132:29–53

Chakraborty PP, Banerjee S, Das NG, Sarkar S, Bose PK (1996) Volcaniclastics and their sedimentological bearing in Proterozoic Kaimur and Rewa Groups in Central India. In: Bhattacharyya A (ed) Recent Advances in Vindhyan Geology, vol 36. Mem Geol Soc India, 59–75

Chakraborty PP, Sarkar S, Bose PK (1998) A viewpoint on intracratonic chenier evolution: clue from a reappraisal of the Proterozoic Ganurgarh Shale, central India. In: Paliwal BS (ed) The Indian Precambrian. Scientific Publishers, Jodhpur, pp 61–72

Chakraborty PP, Sarkar S, Patranabis Deb S (2012) Tectonics and sedimentation of Proterozoic basins of Peninsular India. Proc Ind Nat Sci Acad 78:393–400

Chatterjee BK, Sen PK (1988) Spectral analysis of a Precambrian limestone-shale sequence, Lower Vindhyan, India. Precam Res 39:139–149

Chalapathi Rao NV (2005) A petrological and geochemical reappraisal of the Mesoproterozoic diamondiferous Majhgawan pipe of central India: evidence for transitional kimberlite–orangeite (Group II kimberlite)–lamproite rock type. Mineral Petrol 84:69–106

Chalapathi Rao NV (2006) Mesoproterozoic diamondiferous ultramafic pipes at Majhgawan and Hinota, Panna area, central India: Key to the nature of sub–continental lithospheric mantle beneath the Vindhyan Basin. J Earth Sys Sci 115:161–183

Chanda SK, Bhattacharyya A (1982) Vindhyan sedimentation and palaeogeography: post–Auden developments. In: Valdiya KS, Bhatia SB, Gaur VK (eds) Geology of Vindhyanchal. Hindustan Publ Corp, New Delhi, pp 88–101

Chapmann F (1935) Primitive fossils, possibly Atrematous and Neotrematous brachiopod from the Vindhyans of India. Rec Geol Surv India 69:109–120

Chaudhuri AK, Chanda SK (1991) The Proterozoic basin of the Pranhita–Godavari valley: an overview. In: Tandon SK, Pant CC, Casshyap SM (eds) Sedimentary Basins of India, Tectonic Context. Gyanodayan Prakashan, Nainital, pp 13–29

Choubey VD (1971) Narmada–Son Lineament. Nature 232:38–40

Cloud P (1976) Beginnings of biospheric evolution and their biogeochemical consequences. Paleobiology 2:351–387

Conway Morris S (2000) The Cambrian "explosion": slow–fuse or megatonnage? Proc Nat Acad Sci USA 97:4426–4429

Conway Morris S (2003) The Cambrian "explosion" of metazoans and molecular biology: would Darwin be satisfied? Int J Dev Biol 47:505–515

Crawford AR (1978) Narmada–Son lineament of India traced into Madagascar. J Geol Soc India 19:14–153

Crawford AR, Compston W (1970) The age of Vindhyan system of peninsular India. J Geol Soc Lond 125:351–371

Das AK, Baruah RM, Bisht SS, Agrawal B (1999) An integrated analysis of late Proterozoic Lower Vindhyan Sediments for hydrocarbon exploration in western part of Son valley, central India. J Geol Soc India 53:239–253

Das LK, Misra DC, Ghosh D, Banerjee B (1990) Geomorphotectonics of the basement in a part of Upper Son Valley of the Vindhyan basin. J Geol Soc India 35:445–458

Davaud E, Strasser A, Jedoui Y (1994) Stromatolite and serpulid bioherms in a Holocene restricted lagoon (Sabkha el Melah, southeastern Tunisia). In: Bertrand-Sarfati J, Monty C (eds) Phanerozoic Stromatolites II. Kluwer Acad Publ, Dordrecht, pp 131–151

De C (2003) Possible organisms similar to Ediacaran forms from the Bhander Group, Vindhyan Supergroup, late Neoproterozoic of India. J Asian Earth Sci 21:387–395

De C (2006) Ediacara fossil assemblage in the Upper Vindhyans of central India and its significance. J Asian Earth Sci 27:660–683

Du R, Tian L (1985) Algal macrofossils from the Qingbeikou system in the Yanshan range of North China. Precam Res 29:5–14

Duan C (1982) Late Precambrian algal megafossils Chuaria and Tawuia in some areas of eastern China. Alcheringa 6:57–68

Dutta S, Steiner M, Banerjee S, Erdtmann BD, Jeevankumar S, Mann U (2006) Chuaria circularis from the early Mesoproterozoic Suket Shale, Vindhyan Supergroup, India: insights from light and electron microscopy and pyrolysis–gas chromatography fossils from the Vindhyan Supergroup. J Earth Sys Sci 115:99–112

Eriksson PG, Mazumder R, Sarkar S, Bose PK, Alterman W, Vander-Merwee R (1999) The 2.7–2 Ga volcano–sedimentary record of Africa, India and Australia: Evidence for global and local changes in sea–level and continental freeboard. Precam Res 97:269–302

Eriksson PG, Sarkar S, Banerjee S, Porada H, Catuneanu O, Samanta P (2010) Paleoenvironmental context of microbial mat–related structures in siliciclastic rocks: examples from the Proterozoic of India and South Africa. In: Seckbach J, Oren A (eds) Microbial Mats: Modern and Ancient Microorganisms in Stratified Systems. Springer, Berlin, p 73

Evitt WR (1963) A discussion and proposals concerning fossil dinoflagellates, hystrichospheres and acritarchs. Proc Nat Acad Sci USA 49:158–164

Fedonkin MA (2003) The origin of the Metazoa in the light of the Proterozoic fossil record. Pal Res 7:9–41

Ford TD, Breed WJ (1973) The problematical Precambrian fossil Chuaria. Palaeontology 16:535–550

Ghosh SK (1971) Petrology of Porcellanite rocks of the Samaria area, Sidhi District, Madhya Pradesh. Quart J Geol Min Met Soc India 43:153–164

Gilleaudeau GJ, Sahoo SK, Kah LC, Henderson MA, Kaufman AJ (2018) Proterozoic carbonates of the Vindhyan Basin, India: Chemostratigraphy and diagenesis. Gondwana Res 57:10 25

Gopalan K, Kumar A, Kumar S, Vijayagopal B (2013) Depositional history of the Upper Vindhyan succession, central India: time constraints from Pb–Pb isochron ages of its carbonate components. Precam Res 233:108–117

Gregory LC, Meert JG, Pradhan V, Pandit MK, Tamrat E, Malone SJ (2006) A paleomagnetic and geochronologic study of the Majhgawan kimberlite, India: implications for the age of the Upper Vindhyan Supergroup. Precam Res 149:65–75

Halder D, Ghosh DB (1978) Tectonics of kimberlites around Majhgawan, MP, India. Geol Surv India Misc Publ 34:1–13

Han TM, Runnegar B (1992) Megascopic eukaryotic algae from the 2.1 billion–year old Negaunee Iron-Formation, Michigan. Science 257:232–235

Heron AM (1953) The geology of central Rajputana. Mem Geol Surv India 79:1–389

Hofmann HJ (1971) Precambrian fossils, pseudofossils and problematica in Canada. Bull Geol Surv Canada 189:1–146

Hofmann HJ, Chen J (1981) Carbonaceous megafossils from the Precambrian (1800 Ma) near Jixian Northern China. Can J Ear Sci 18:443–447

Jones HJ (1909) In: General Report, vol 3. Rec Geol Surv India, 66

Kaila KL, Murthy PRK, Mall DM (1989) The evolution of the Vindhyan Basin vis–à–vis the Narmada–Son lineament, Central India, from deep seismic soundings. Tectonophys 162:277–289

Kaila KL, Reddy PR, Dixit MM, Rao PK (1985) Crustal structure across the Narmada–Son lineament, central India, from deep seismic soundings. J Geol Soc India 26:465–480

Kailasam LN (1979) Plateau uplift in Peninsular India. Tectonophys 61:243–269

Kathal PK, Patel DR, Alexander PO (2000) An Ediacaran fossil Spriggina (?) from the Semri Group and its implication on the age of the Proterozoic Vindhyan Basin, Central India. N. Jb. Geol Palaont Mon 6:321–332

Knoll AH (2003) Vestiges of a beginning? Paleontological and geochemical constraints on early animal evolution. Annales de Pal 89:205–221

Krishnan MS, Swaminath J (1959) The great Vindhyan Basin of Northern India. J Geol Soc India 1:10–30

Kulkarni KG, Borkar VD (1996a) A significant stage of metazoan evolution from the Proterozoic rocks of the Vindhyan Supergroup. Curr Sci 70:1096–1097

Kulkarni KG, Borkar VD (1996b) Occurrence of Cochlichnus hitchcock in the Vindhyan Supergroup (Proterozoic) of Madhya Pradesh. J Geol Soc India 47:725–729

Kumar S (1976a) Stromatolites from the Vindhyan rocks of the Son Valley-Maihar area, district Mirzapur (UP) and Satna (MP). J Palaeontol Soc India 18:13–21

Kumar S (1976b) Significance of stromatolites in the correlation of Semri Series (Lower Vindhyan) of Son Valley and Chitrakut area. UP J Palaeontol Soc India 19:24–27

Kumar S (1978) Discovery of microorganisms from the black cherts of the Fawn limestone (Late Precambrian) Semri Group, Son Valley, Mirzapur district, UP. Curr Sci 47:461

Kumar S (1980) Stromatolites and Indian biostratigraphy. J Palaeotol Soc India 23:166–183

Kumar S (1982) Vindhyan stromatolites and their stratigraphic testimony. In: Valdia KS, Bhatia SB, Gaur VK (eds) Geology of Vindhyanchal, pp 102–112

Kumar S (1984) Present status of stromatolites biostratigraphy in India. Geophytology 14:96–110

Kumar S (1995) Megafossils from the Mesoproterozoic Rohtas Formation (the Vindhyan Supergroup), Katni area, central India. Precam Res 72:171–184

Kumar S (2001) Mesoproterozoic megafossil Chuaria–Tawuia association may represent parts of a multicellular plant, Vindhyan Supergroup, central India. Precam Res 106:187–211

Kumar S (2016) Megafossils from the Vindhyan Basin, central India: an overview. J Palaeontol Soc India 61:273–286

Kumar S, Gupta S (2002) International Field Workshop on the Vindhyan Basin, central India. Field Guide Book, Pal Soc India, Lucknow

Kumar S, Srivastava P (1995) Microfossils from the Kheinjua Formation, Mesoproterozoic Semri Group, Newari area, central India. Precam Res 74:91–117

Kumar S, Srivastava P (1997) A note on the carbonaceous megafossils from the Neoproterozoic Bhander Group, Maihar area, Madhya Pradesh. J Palaeontol Soc India 42:141–146

Kumar S, Schidlowski M, Joachimski MM (2005) Carbon isotope stratigraphy of the Paleo–Neoproterozoic Vindhyan Supergroup, central India: implications for basin evolution and intrabasinal correlation. J Palaeontol Soc India 50:65–81

Logan BH (1961) Cryptozoon and associate stromatolites from the Recent, Shark Bay, Western Australia. J Geol 69:517–533

Maithy PK (1991) On *Krishnania* Sahni and Shrivastava, a Mid-Proterozoic macrofossil. J Palaeontol Soc India 36:59–65

Maithy PK, Shukla M (1977) Microbiota from the Suket Shales, Rampura, Vindhyan System, Madhya Pradesh. Palaeobotan 23:176–188

Maithy PK, Mandal J (1983) Microbiota from Vindhyan Supergroup of Karauli-Sapotra region of northeast Rajasthan, India. Palaeobolan 31:129–142

Maithy PK, Shukla M (1984) Biological remains from the Suket Shale formation, Vindhyan Supergroup. Geophytol 14:212–215

Maithy PK, Gupta S (1983) Biota and organosedimentary structures from the Vindhyan Supergroup around Chandrehi, Madhya Pradesh. Palaeobotan 31:154–164

Maithy PK, Babu R (1988) The mid–Proterozoic Vindhyan Microbiota from Chopan, Southeast Uttar Pradesh. J Geol Soc India 31:584–590

Maithy PK, Meena KL (1989) Organic-walled microfossils from the Proterozoic succession of Vindhyan Supergroup exposed around Satna, Madhya Pradesh, India. Indian J Earth Sci 16:178–188

Mallet FR (1869) On the Vindhyan Series exhibited in northwestern and central Provinces of India. Geol Surv India Mem 7:1–129

Malone SJ, Meert JG, Banerjee DM, Pandit MK, Tamrat E, Kamenov GD, Pradhan VR, Sohl LE (2008) Paleomagnetism and detrital zircon geochronology of the Upper Vindhyan sequence, Son Valley and Rajasthan, India: a ca. 1000 Ma closure age for the Purana basins? Precam Res 164:137–159

Mandal S, Choudhuri A, Mondal I, Sarkar S, Chakraborty PP, Banerjee S (2019) Revisiting the boundary between lower and upper Vindhyan, Son valley, India. J Earth Sys Sci, doi.10.1007/s12040-019-1250-2

Mathur SM (1982) Organic materials in the Precambrian Vindhyan Supergroup. In: Valdiya KS, Bhatia SB, Gaur VK (eds) Geology of Vindhyanchal. Hindustan Publ Corp, India, pp 125–131

Mathur SM (1983) A new collection of fossils from the Precambrian Vindhyan Supergroup of central India. Curr Sci 52:363–365

Mathur SM, Singh HN (1971) Petrology of the Majhgawan pipe rock. Geol Surv India Misc Publ 19:78–85

Mazumder R, Bose PK, Sarkar S (2000) A commentary on the tectonic–sedimentary record of the Pre–2.0 Ga evolution of Indian craton vis–à–vis Pre-Gondwana Afro–Indian Supercontinent. J African Earth Sci 30:201–217

McMenamin OS, Kumar S, Awramik SM (1983) Microbial fossils from the Kheinjua Formation, Middle Proterozoic, Semri Group (Lower Vindhyan), Son Valley area, central India. Precam Res 34:247–272

Misra RC (1957) Fermoria, the enigma of Indian palaeontology. J Palaeontol Soc India 2:54–57

Misra RC, Bhatnagar GS (1950) On carbonaceous discs and 'algal dust' from the Vindhyan Pre-Cambrian. Curr Sci 19:88–89

Misra RC, Dube SN (1952) A new collection and restudy of the organic remains from the Suket shales (Vindhyans) Rampura, Madhya Bharat. Sci Cult 18:46–48

Misra RC, Awasthi N (1962) Sedimentary markings and other structures in the rocks of Vindhyan formation of the Son valley and Maihar–Rewa area, India. J Sed Petrol 32:764–775

Mishra M, Bickford ME, Basu A (2018) U–Pb age and chemical composition of an ash bed in the Chopan Porcellanite formation, Vindhyan Supergroup, India. J Geol 126:553–560

Nair KKK, Jain SC, Yedekar DB (1995) Stratigraphy, structure and geochemistry of the Mahakoshal greenstone belt. Mem Geol Soc India 31:403–432

Naqvi SM, Rogers JJW (1987) Precambrian Geology of India. Clarendron Press, Oxford

Narain H (1987) Geophysical constraints on the evolution of Purana basins of India with special reference to Cuddapah, Godavari and Vindhyan basins. Mem Geol Soc India 6:5–12

Nautiyal AC (1983) Algonkian (Upper to Middle) micro-organisms from the Semri Group of Son Valley (Mirzapur District), India. Geoscience 4:169–198

Nautiyal AC (1986) Lower Vindhyan (Algonkian) microflora (microbiota) and biostratigraphy of Sangrampur hill, Banda district, Northern India. Geos J 7:1–22

Oldham T (1856) Remarks on the classification of the rocks of central India resulting from the investigation of the Geological Survey. J Asiatic Soc Bengal 25:224–256

Paikaray S, Banerjee S, Mukherjee S (2008) Geochemistry of shales from the Paleoproterozoic to Neoproterozoic Vindhyan Supergroup: implications on provenance, tectonics and paleoweathering. J Asian Ear Sci 32:34–48

Paul DK (1991) Indian kimberlites and lamprophyres: mineralogical and chemical aspects. J Geol Soc India 37:221–238

Paul DK, Rex DC, Harris PG (1975) Chemical characteristics and K–Ar ages of Indian kimberlites. Bull Geol Soc Am 86:364–366

Peterson KJ, McPeek MA, Evans DAD (2005) Tempo and mode of early animal evolution: inferences from rocks, hox and molecular clocks. Paleobiol 31:36–55

Prakash R, Dalela K (1982) Stratigraphy of the Vindhyan in Uttar Pradesh: a brief review. In: Valdiya KS, Bhatia SB, Gaur VK (eds) Geology of Vindhyanchal. Hindustan Publ Corporation, New Delhi, pp 55–79

Prasad B (1980) Vindhyan stromatolite biostratigraphy. Geol Surv India Misc Publ 44:201–2016

Prasad B (1984) Geology, sedimentation and paleogeography of the Vindhyan Supergroup, SE Rajasthan. Mem Geol Surv India 116:1–107

Prasad B, Ramaswamy SM (1980) Stromatolites in Upper Vindhyan from Bundi, Kota and Sawai Madhopur District, Rajasthan: Stromatolites characteristic and utility. Geol Surv India Misc Publ 44:275–277

Prasad B, Verma KK (1991) Vindhyan Basin—a review. In: Tandon SK, Pant CC, Casshyap SM (eds) Sedimentary Basins of India: TectonicCcontext. Gyanodaya Prakashan, Nainital, pp 50–62

Raaben ME (1969) Columnar stromatolites and late Precambrian stratigraphy. Am J Sci 261:1–18

Raaben ME (2005) On subdivisions of the Upper Riphean. Strat Geol Corr 13:143–158

Raha PK, Sastri MVA (1982) Stromatolites and Precambrian stratigraphy in India. Precam Res 18:292–318

Rai V, Gautam R (1998) New occurrence of carbonaceous megafossils from the Meso- to Neoproterozoic horizons of the Vindhya Supergroup, Kaimur–Katni area, Madhya Pradesh, India. Geophytol 26:13–25

Rai V, Shukla M, Gautam R (1997) Discovery of carbonaceous megafossils (*Chuaria–Tawuia* assemblage) from the Neoproterozoic Vindhyan succession (Rewa Group), Allahabad–Rewa area, India. Curr Sci 73:783–788

Ram J (2005) Hydrocarbon exploration in onland frontier basins of India–prospectives and challenges. J Pal Soc India 50:1–16

Ram J, Shukla SN, Pramanik AG, Verma BK, Chandra G, Murthy MSN (1996) Recent investigations in the Vindhyan Basin: implications for the basin tectonics. In: Bhattacharyya A (ed) Recent Advances in Vindhyan Geology, vol 36. Mem Geol Soc India, pp 267–286

Rao KS, Lal C, Ghosh DB (1977) Algal stromatolites in the Bhander Group, Vindhyan Supergroup, Satna district, Madhya Pradesh. Rec Geol Surv India 109:38–47

Rao KS, Neelakantam S (1978) Stratigraphy and sedimentation of Vindhyans in parts of Son valley area, Madhya Pradesh. Rec Geol Surv India 110:180–193

Rasmussen B, Bose PK, Sarkar S, Banerjee S, Fletcher IR, McNaughton NJ (2002) 1.6 Ga U–Pb zircon age for the Chorhat Sandstone, lower Vindhyan, India: possible implications for early evolution of animals. Geology 30:103–106

Rasmussen B, Fletcher IR, Bengtson S, McNaughton NJ (2004) SHRIMP U–Pb dating of diagenetic xenotime in the Stirling Range Formation, Western Australia: 1.8–billion–year minimum age for the Stirling biota. Precam Res 133:329–337

Shanker Ravi, Nag S, Ganguly A, Absar A, Rawat BP, Singh GS (2001) Are Majhgawan–Hinota pipe rocks truly Group—I kimberlite? Proc Indian Acad Sci (Earth Planet Sci) 110:63–76

Ray JS, Martin MW, Veizer J, Bowring SA (2002) U–Pb zircon dating and Sr isotope systematics of the Vindhyan Supergroup, India. Geology 30:131–134

Ray JS, Veizer J, Davis WJ (2003) C, O, Sr and Pb isotope systematics of carbonate sequences of the Vindhyan Supergroup, India: age, diagenesis, correlations and implications for global events. Precam Res 121:103–140

Ray JS (2006) Age of the Vindhyan Supergroup: a review of recent findings. J Earth Sys Sci 115:149–160

Raza M, Casshyap SM (1996) A tectonic-sedimentary model of evolution of middle Proterozoic Vindhyan Basin. In: Bhattacharyya A (ed) Recent Advances in Vindhyan Geology, vol 36. Mem Geol Soc India, pp 286–300

Raza M, Casshyap SM, Khan A (2001) Accretionary lapilli from the Basal Vindhyan Volcanic sequence, south of Chittaurgarh, Rajasthan and their implication. J Geol Soc India 57:77–82

Rogers JJW (1986) The Dharwar craton and assembly of peninsular India. J Geol 94:129–144

Roy A, Bandyopadhyay BK (1990) Tectonics and structural pattern of the Mahakoshal belt of central India. Spec Publ Geol Surv India 28:226–240

Sahni MR (1936) *Fermeria minima*: a revised classification of the organic remains from the Vindhyan of India. Rec Geol Surv India 69:458–468

Sarkar B (1974) Biogenic sedimentary structures and microfossils of the Bhander Limestone (Proterozoic) in India. Quart J Geol Min Metal Soc India 46:143–156

Sarangi S, Gopalan K, Kumar S (2004) Pb–Pb age of earliest megascopic, eukaryotic alga bearing Rohtas Formation, Vindhyan Supergroup, India: implications for Precambrian atmospheric oxygen evolution. Precam Res 132:107–121

Sarkar S, Banerjee S (2007) Some unusual and/or problematic features. In: Schieber J, Bose PK, Eriksson PG, Banerjee S, Sarkar S, Catuneanu O, Altermann W (eds) An Atlas of Microbial Mat Features Preserved Within the Clastic Rock Record. Elsevier, Amsterdam, pp 145–147

Sarkar S, Banerjee S, Bose PK (1996) Trace fossils in the Mesoproterozoic Koldaha Shale, central India, and their implications. N Jb fur Geol und Palaontologie-Monatshefte 7:425–438

Sarkar S, Chakraborty PP, Bhattacharya SK, Banerjee S (1998) C12–enrichment along intraformational unconformities within Proterozoic Bhander limestone, Son valley, India and its implication. Carb Evap 13:108–114

Sarkar S, Banerjee S, Chakraborty S, Bose PK (2002a) Shelf storm flow dynamics: insight from the Mesoproterozoic Rampur Shale, central India. Sed Geol 147:89–104

Sarkar S, Chakraborty S, Banerjee S, Bose PK (2002b) Facies sequence and cryptic imprint of sag tectonics in late Proterozoic Sirbu Shale, central India. In: Altermann W, Corcoran, P (eds) Precambrian Sedimentary Environments: a Modern Approach to Ancient Depositional Systems, vol 33. Spec Publ Int Ass Sedimentol, Blackwell Science, pp 369–382

Sarkar S, Banerjee S, Eriksson PG (2004) Microbial mat features in sandstone illustrated. In: Eriksson PG, Altermann W, Nelson W, Mueller DR, Catuneanu O (eds) The Precambrian Earth: Tempos and Events. Elsevier, Amsterdam, pp 673–675

Sarkar S, Banerjee S, Eriksson PG, Catuneanu O (2005) Microbial mat control on siliciclastic Precambrian sequence stratigraphic architecture: examples from India. Sed Geol 176:195–209

Sarkar S, Banerjee S, Samanta P, Jeevankumar S (2006) Microbial mat–induced sedimentary structures and their implications: examples from Chorhat Sandstone, MP, India. J Earth Sys Sci 115:49–60

Sarkar S, Choudhuri A, Banerjee S, van Loon AJ, Bose PK (2014a) Seismic and non-seismic soft-sediment deformation structures in the Proterozoic Bhander Limestone, central India. Geologos 20:89–103

Sarkar S, Banerjee S, Samanta P, Chakraborty N, Mukhopadhyay S, Chakraborty P, Singh A (2014b) Microbial mat records in siliciclastic rocks: examples from Four Indian Proterozoic basins and their modern equivalents in Gulf of Cambay. J Asian Ear Sci 91:362–377

Sarkar S, Choudhuri A, Mandal S, Eriksson PG (2016) Microbial mat related structures shared by both siliciclastic and carbonate formations. J Palaeogeography 5:278–291

Schieber J (1998) Possible indicators of microbial mat deposits in shales and sandstones: examples from the Mid-Proterozoic Belt Supergroup, Montana, USA. Sed Geol 120:105–124

Schieber J (2004) Microbial mats in the siliciclastic rock record: a summary of the diagnostic features. In: Eriksson PG, Altermann W, Nelson DR, Muller WU, Catuneanu O (eds) The Precambrian Earth: Tempos and Events. Elsevier, Amsterdam, pp 663–673

Schieber J, Bose PK, Eriksson PG, Banerjee S, Sarkar S, Catuneanu O, Altermann W (2007) An Atlas of Microbial Mat Features Preserved Within the Siliciclastic Rock Record. Elsevier Science, Amsterdam, p 311p

Schopf JW (1975) Precambrian palaeobiology: problems and perspectives. Ann Rev Ear Planet Sci 3:213–249

Schopf W (2004) Earth's earliest biosphere: status of the hunt. In: Eriksson PG, Altermann W, Nelson DR, Muller WU, Catuneanu O (eds) The Precambrian Earth: Tempos and Events. Elsevier, Amsterdam, pp 516–539

Schopf JW, Walter MR (1983) Archean microfossils: new evidences of ancient microbes. In: Schopf JM (ed) Earth's Earliest Biosphere. Princeton University Press, Princeton, New Jersey, pp 214–239

Seilacher A (1997) Fossil Art. The royal tyrell museum of paleontology, Drumheller, Canada

Seilacher A (2007) Trace Fossil Analysis. Springer, Berlin

Seilacher A, Bose PK, Pflüger F (1998) Triploblastic animals more than 1 billion years ago: trace fossil evidence from India. Science 282:80–83

Sharma M (1996) Microbiolites (stromatolites) from the Mesoproterozoic Salkhan Limestone, Semri Group, Rohtas, Bihar: their systematics and significance. Geol Soc India Mem 36:167–196

Sharma M, Sergeev VN (2004) Genesis of carbonate precipitate patterns and associated microfossils in Mesoproterozoic formations of India and Russia—a comparative study. Precam Res 134:317–347

Shukla M, Sharma M (1990) Palaeobiology of Suket Shale, Vindhyan Supergroup; age implications. Geol Surv India Spec Publ 28:411–434

Sharma M, Mishra S, Dutta S, Banerje S, Shukla Y (2009) On the affinity of *Chuaria–Tawuia* complex: a multidisciplinary study. Precam Res 173:123–136

Sharma M, Kumar S, Tiwari M, Shukla Y, Pandey SK, Srivastava P, Banerjee S (2012) Palaeobiological constraints and the Precambrian biosphere: Indian evidence. Proc Nat Sci Acad 78:407–422

Singh HJM, Banerjee AK (1980) Stromatolites in the Vindhyan Group of Northeastern Rajasthan. Misc Publ Geol Surv India 44:278–283

Singh IB (1973) Depositional environment of the Vindhyan sediments in the Son valley area. Recent Researches in Geology, vol 1. Hindusthan Publ Corp, New Delhi, pp 140–152

Singh IB (1980) Precambrian sedimentary sequences of India: their peculiarities and comparison with modern sediments. Precam Res 12:411–436

Singh IB (1985) Paleogeography of the Vindhyan Basin and its relationship with late Proterozoic basins of India. J Palaeontol Soc India 30:35–41

Singh SP, Sinha PK (2001) Vindhyan Supergroup of Bihar–an overview. In: Singh SP (ed) Precambrian Crustal Evolution and Metallogeny of India. South Asian Ass of Econ Geol Patna, pp 107–126

Singh SP, Thakur LK, Sinha AK (2001) Stratigraphy of the Semri Group in the Bhaunathpur area, Garwah District, Jharkhand. In: Singh SP (ed) Precambrian Crustal Evolution and Metallogeny of India. South Asian Ass of Econ Geol Patna, Patna, pp 95–106

Soni MK, Chakraborty S, Jain VK (1987) Vindhyan Supergroup—a review. In: Purana basins of peninsular India (Middle to Late Proterozoic), vol 6. Mem Geol Soc India, pp 87–138

Srivastava DC, Sahay A (2003) Brittle tectonics and pore–fluid conditions in the evolution of the Great Boundary Fault around Chittaurgarh, Northwestern India. J Struc Geol 25:1713–1733

Srivastava RN (1971) Microorganic remains from the Vindhyan formations of India. In: Proceeding of the seminar on palaeopalynology and Indian stratigraphy, Calcutta, 1–14

Srivastava RN (1977) Environmental significance of some depositional structures in banded porcellanites (Lower Vindhyan) of Mirzapur District, UP. J Indian Ass Sediment 1:45–51

Srivastava JP, Iqbaluddin (1981) Some recent observation in Son valley, Mirzapur district, U.P. Geol Surv India Misc Pub 50: 99–108

Srivastava AP, Rajagopalan G (1988) F-T ages of Vindhyan glauconitic sandstone beds exposed around Rawatbhata area, Rajasthan. J Geol Soc India 32:527–529

Srivastava P (2004) Carbonaceous fossils from the Panna Shale, Rewa Group (Upper Vindhyans), central India: a possible link between evolution of micro-megascopic life. Curr Sci 86:644–646

Steiner M (1997) Chuaria circularis WALCOTT 1899—"Megasphaeromorph acritarch" or prokaryotic colony? Acta Univ Carol Geol 40:645–665

Sun W (1987) Palaeontology and biostratigraphy of late Precambrian macroscopic colonial algae: Chuaria and Tawuia Hofmann. Palaeontographica Abt 203:109–134

Sur S, Schieber S, Banerjee S (2006) Petrographic observations suggestive of microbial mats from Rampur Shale and Bijaigarh Shale, Vindhyan Basin, India. J Earth Sys Sci 115:61–66

Tandon KK, Kumar S (1977) Discovery of annelid and arthropod remains from Lower Vindhyan rocks (Precambrian) of central India. Geophytol 7:126–128

Tewari AP (1968) A new concept of the paleotectonic set-up of a part of northern peninsular India with special reference to the Great Boundary Faults. Geol en Mijnb 47:21–27

Tripathy GR, Singh SK (2015) Re–Os depositional age for black shales from the Kaimur Group, Upper Vindhyan, India. Chem Geol 413:63–72

Turner CC, Meert JG, Pandit MK, Kamenov GD (2014) A detrital zircon U–Pb and Hf isotopic transect across the Son Valley sector of the Vindhyan Basin, India: implications for basin evolution and paleogeography. Gond Res 26:348–364

Valdiya KS (1969) Stromatolites of the lesser Himalayan Carbonate formations and the Vindhyan. J Geol Soc India 10:1–125

Valdiya KS (1982) Tectonic perspectives of the Vindhyachal region. In: Valdiya KS, Bhatia SB, Gaur VK (eds) Geology of Vindhyanchal. Hindustan Publishing Corporation, New Delhi, pp 23–29

Venkateshwarlu M, Rao JM (2013) Paleomagnetism of Bhander sediments from Bhopal Inlier, Vindhyan Supergroup. J Earth Sys Sci, 330–336

Venkatachala BS, Sharma M, Shukla M (1996) Age and life of Vindhyans: facts and conjectures. In: Bhattacharyya A (ed) Recent Advances in Vindhyan Geology, vol 36. Mem Geol Soc India, 137–166

Verma PK (1991) Geodynamics of the Indian Peninsula and the Indian Plate Margin. Oxford and IBH, 357

Verma PK, Banerjee P (1992) Nature of continental crust along the Narmada–Son Lineament inferred from gravity and deep seismic sounding data. Tectonophys 202:375–397

Vinogradov AP, Tugarinov AI, Zhikov CI, Stanikova NI, Bibikova EV, Khorre K (1964) Geochronology of the Indian Precambrian. In: Report 22nd international geological congress, New Delhi, vol 10, 553–567

Wang DYC, Kumar S, Hedges SB (1999) Divergence time estimates for the early history of animal phyla and the origin of plants, animals and fungi. Proc Royal Soc Lond Bio Sci 266:63–171

West WD (1962) The line of the Narmada and Son valleys. Cur Sci 31:143–144

Williams GE, Schmidt PW (2003) Possible fossil impression in sandstone from the late Paleoproterozoic-early Mesoproterozoic Semri Group (lower Vindhyan Supergroup), central India. Alcheringa 27:75–76

Wray GA, Levinton JS, Shapiro LH (1996) Molecular evidence for deep Precambrian divergences among metazoan phyla. Science 274:568–581

Zhu S, Sun S, Huang X, Zhu G, Sun L, Kuan Z (2000) Discovery of carbonaceous compressions and their multicellular tissues from the Changzhougou Formation (1800 Ma) in the Yanshan range, North China. Chi Sci Bull 45:841–847

Chapter 2
Facies, Paleogeography and Sequence Stratigraphy

2.1 Introduction

The Vindhyan rocks exposed in the Son valley unconformably overlies the granites/metasediments of the Mahakoshal Group. The Supergroup with its excellent preservation records the interplay between sediments and different sedimentological processes. Excellent preservation of primary sedimentary structures allows a detailed process-based facies analysis. The following section deals with the detailed facies descriptions and interpretations of the formations constituting the Vindhyan Supergroup, followed by paleogeographic interpretations. Facies of the constituent formations and members are presented in tables. The sequence stratigraphic framework of the Vindhyan Supergroup has been summarized at the end. Coordinates of the places are provided at the end of this chapter.

2.2 Deoland Formation

The Deoland Formation, which is underlain by Archean basement and succeeded by the dark green-coloured Arangi Shale above, is an overall fining-upward siliciclastic succession (Banerjee et al. 2008). Locally, the Formation may exhibit a conglomerate bed of ca 2.8 m in thickness, considered as a glacier deposit (Dubey and Chaudhary 1952; Chaudhary 1953; Ahmad 1955, 1958; Mathur 1954, 1960, 1981), although differing views exist (Williams and Schmidt 1996). The overlying part of the Deoland Formation, which originated in the inner shelf, gradationally passes over to the Arangi Shale that comprises the lower part of the Kajrahat Formation. The Deoland Formation is well exposed around Sidhi, Shikarganj and Chopan area. A detailed facies analysis around Chopan area reveals two broad facies association within the Deoland Formation (Table 2.1, Fig. 2.1; for details see Banerjee et al. 2008; Banerjee 2010). A thin and broadly wavy granular sheet divides the formation

© Springer Nature Singapore Pte Ltd. 2020
S. Sarkar and S. Banerjee, *A Synthesis of Depositional Sequence
of the Proterozoic Vindhyan Supergroup in Son Valley*, Springer Geology,
https://doi.org/10.1007/978-981-32-9551-3_2

Table 2.1 Description and interpretation of facies constituting the Deoland Formation around Chopan area

Facies	Description	Interpretation
Thinly laminated/graded shale	It consists of thinly laminated grey shale occasionally associated with lighter coloured mm-thick siltstone stringers. Grading is locally discernable within the siltstone stringers. The siltstone bed tops locally bear minute ripples (average wavelength 2 cm, amplitude 0.5 cm)	Shale represents low-energy suspension fallout. Grading within the siltstone stringers suggests deposition from steady waning flows associated with high energy flows
Massive/planar-laminated siltstone	It consists of cm- to decimeter-thick siltstone beds alternating with shales towards the top of the section. Most siltstone beds exhibit sharp bases and gradational tops. They may be entirely massive or may contain planar laminae. All beds are topped by wave ripples with average width 4 cm and amplitude 1.5 cm. The bed soles often bear erosional marks like gutter casts and prod marks	The siltstone beds containing gutters casts and tool marks at their bases, encased by shale below and above suggest deposition from episodic comparatively high energy flows, such as storms
Trough cross-stratified sandstone	It is made of coarse sands, internally characterized by trough cross-stratification. The beds are up to 80 cm thick, lenticular and laterally discontinuous. Thickness of the cross-sets is up to 25 cm. Cross-strata azimuth provides northward paleocurrent direction. The facies is closely associated with chevron cross-stratified sandstone facies	Formed by arcuate dune migration possibly in shallow channels
Wavy-laminated sandstone	It shows predominance of sheet-like geometry of sandstone beds with sharp bases and gradational tops. It rests invariably upon the shale facies or the planar-laminated sandstone facies. Gutter casts, prod marks and rip-up mud clasts occur at the base of sandstone beds. Internally, the beds are dominantly wavy-laminated, hummocky with subordinate planar laminae and overall grading	Wavy laminae suggest oscillatory flows. Hummocky cross-stratification indicates combined wave and current actions. Gutter casts, prod marks and rip-up mud clasts indicate supercritical nature of the sand-laden flows
Planar-laminated sandstone	This facies consists of planar-laminated fine sandstone beds, a few centimetres to 85 cm thick, laterally extensive and tabular in geometry. The facies is closely associated with chevron cross-stratified sandstone, hummocky cross-stratified sandstone and shale facies. Every lamina within the beds bear well-pronounced parting lineation	The characteristic planar laminae and associated parting lineation indicate deposition in a high energy (upper flow regime plane bed) environment

(continued)

Table 2.1 (continued)

Facies	Description	Interpretation
Chevron cross-stratified sandstone	This facies is characterized by intrabed criss-cross arrangement of oppositely oriented cross-stratification. Almost all the beds are topped by wave ripples with wavelength and amplitude varying from 15 to 25 cm and 6 to 15 cm, respectively. Chevron cross-set thickness, on the other hand, varies from 25 to 35 cm	Chevron cross-stratifications strongly point to wave agitation causing to-and-fro grain movement. The wave ripples on top of almost every cross-stratified bed possibly indicate later reworking by gentler waves
Pebbly sandstone	It overlies the conglomerate facies or the basement directly and dominates the top part of the lower facies assemblage. Individual sandstone beds are lenticular in geometry with erosional concave-up bases, tens of cm to more than a metre thick and stacked vertically; poor sorting and random arrangement of pebbles. Pebbles are almost exclusively of vein quartz, chert and jasper	The pebbly sandstone beds indicate deposition generally from more fluidal flows than breccia and conglomerate facies. Poor sorting of the pebble fraction and its random distribution within the beds possibly suggests flash-flood deposition
Matrix-supported conglomerate	It consists of matrix-supported conglomerate beds measuring up to 50 cm to 1.5 m in thickness. The bed geometry is broadly lenticular, many showing distinct upward convexity, others slight concavity at base. This facies typically is sandwiched between breccia and pebbly sandstone facies	Abraded nature of the clasts indicates substantial transportation. Random clast fabric suggests deposition from sediment gravity flow. Lenticular shape and general upward convexity of the beds identify most conglomerates as lobes of very viscous debris flows
Massive breccia	It occurs at the bottom part of the lower facies assemblage resting directly on the basement. The brecciated beds are massive, wedge-shaped, laterally discontinuous and consist solely of chaotically arranged, very poorly sorted, angular clasts generally ranging in length up to 15 cm. Subordinate conglomerate with abraded clasts intervene the breccia beds locally. The clasts include quartzite, phyllites, schists and jasper	The size and angularity of the clasts indicate very short transport. Composition of the clasts indicates their derivation from the basement rocks in close vicinity. Matching boundaries between adjacent clasts speaks for in situ brecciation most probably under tectonic impact. Breccia appears to be scree deposit at the foot of relatively steep scarps

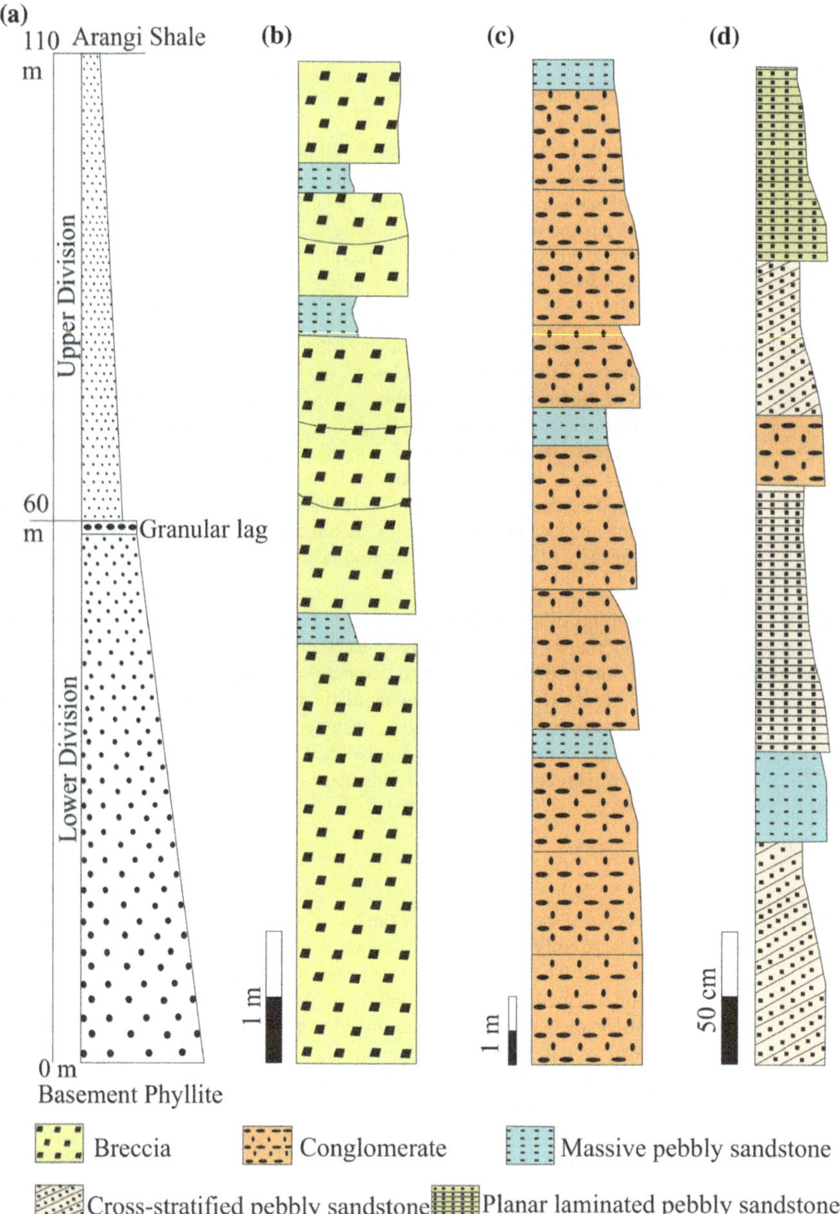

Fig. 2.1 Lithocolumn showing two divisions of the Deoland Formation separated by a granular lag (**a**), and detailed lithologs for breccia facies (**b**), conglomerate facies (**c**) and pebbly sandstone facies (**d**)

into two upward-fining successions, differing in lithology, texture, sand body geometry and internal structures. The lower 60-m-thick segment of the Deoland Formation consists predominantly of coarse-grained sediments. Local patches of fault breccia, with occasional matching boundaries between adjacent clasts, overlie the unconformity on top of the Archean basement rocks. Otherwise, the lower segment is made up of conglomerate and pebbly sandstone, both poorly sorted and possessing lenticular geometry, comparatively more pronounced in the former. The conglomerates are internally massive, but with increasing incorporation of sandstone, become crudely cross-stratified. Although pebbles generally define the foreset bases, these are also randomly scattered. The sandstone is thoroughly cross-stratified, except at the base of the segment where it is massive or poorly cross-stratified. The vertical juxtaposition of channel forms of these poorly sorted siliciclastic sedimentary rocks supports fluvial aggradation. The fining-upward trend in this Lower Deoland stratigraphic segment reflects a slow rise in base level. Sandstone lenses with pebbles scattered randomly within them indicate intermittent high-energy flash-flood deposition (cf. Pfluger and Seilacher 1991).

In contrast, the 50-m-thick Upper Deoland stratigraphic segment is distinctly fine-grained, progressively fines upwards and consists dominantly of chevron cross-stratified sandstone, hummocky cross-stratified sandstone- and planar-laminated sandstone, with occasional trough cross-stratified sandstone (Figs. 2.2, 2.3). The sandstone is well sorted and consists of well-rounded grains. The sandstone beds are tabular or sheet-like in geometry. The soles of the beds locally contain gutter and groove casts, as well as prod marks. The sandstone of the upper segment is selectively glauconitized (Banerjee et al. 2008; Banerjee 2010). Muddy siltstone progressively dominates the upper part of this segment and becomes over-thickened towards the top.

The overall fining-upward upper segment of the Deoland Formation gradationally passes over to the Arangi Shale of the Kajrahat Formation. The abundance of wave-formed features including hummocky cross-strata, wave ripples, chevron cross-strata, quasi-planar strata and low-angle trough cross-strata suggest that the deposition of the upper part of the Deoland Formation took place on an open, strongly agitated shelf, that deepened through time. The granular lag at the base of the segment marks a ravinement surface that represents a substantial increase in the rate of base level or sea level rise. Transgression continued throughout the deposition of the Upper Deoland Formation, to the Arangi shale which is of deep offshore origin (Bose et al. 2001). The overall fining-upward Deoland Formation and the Arangi Shale together form a transgressive systems tract (Banerjee et al. 2008).

2.3 Kajrahat Formation

The Arangi Shale is poorly exposed in the entire Son valley area. In places, it is marked by carbonaceous shale with total organic carbon content exceeding 5% (Banerjee et al. 2006a; Singh et al. 2018). The Arangi Shale gradationally passes

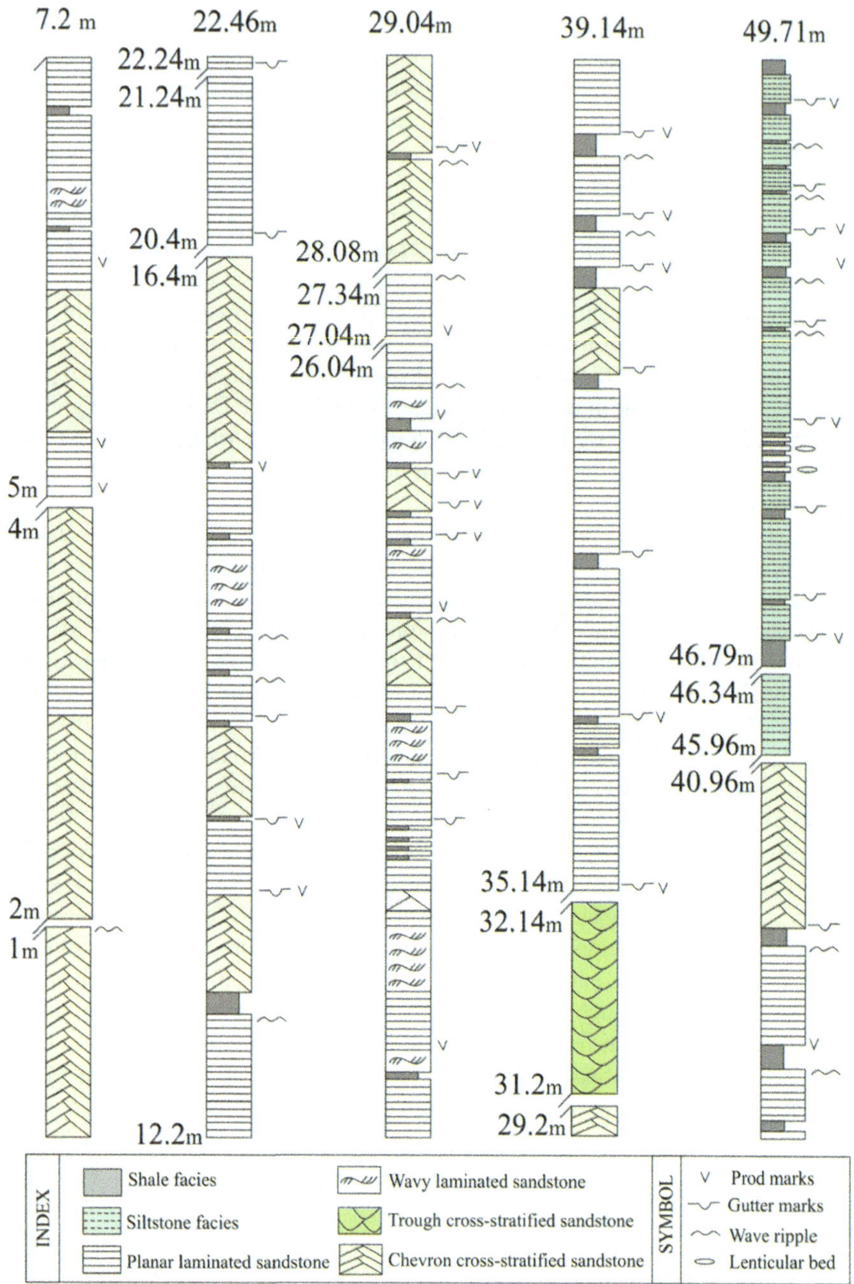

Fig. 2.2 Stratigraphic succession of the upper segment of the Deoland Formation showing facies distribution

Fig. 2.3 Contact between the basement and the Deoland Formation (**a**), poorly sorted pebbly sandstone (**b**), hummocky cross-stratification (**c**) and mega-ripples on sandstone bed surface (**d**) (pen length in b = 14.5 cm, marker length in c = 13.5 cm)

upwards into the Kajrahat Limestone, which is frequently dolomitic and exhibits emergence features at the top part. Various types of stromatolites are present within this limestone (Figs. 2.4, 2.5, 2.6). Good exposures of the Kajrahat Limestone occur around Kuteswar–Dhanwahi area, south of Maihar and around Dala cement factory. The Kajrahat Limestone is nicely exposed in ~250-m-thick section around Kuteswar limestone mines. However, many good exposures are generally drowned underwater in surrounding areas. Sandwiched between the Arangi Shale below and the Porcellanite Formation above, the Kajrahat Limestone at Kuteswar comprises three superposed divisions with distinctive facies assemblages (Fig. 2.4). A detailed description of individual facies and their characteristic features are given in Table 2.2 (for details see Banerjee et al. 2007). The 60-m-thick basal division of the Kajrahat Limestone consists entirely of dolomites. It consists of a grey massive dolomite body (facies A) interspersed with isolated yellowish-grey, planar-curved, cross-stratified dolomite lenses (facies B). The 70-m-thick middle division consists of monotonous vertical alternations of dark grey, faintly laminated limestone (facies C) and yellowish-grey dolomite (facies D, Fig. 2.4). Barring a few lensoid and cross-stratified dolomite bodies the 125-m-thick upper division is composed of stromatolites and microbial mat laminae (Fig. 2.5). These microbial facies consist of non-ferroan calcimicrite, recrystallized to non-ferroan calcite micro-spars. The stromatolites occur as two different forms, as large (facies E) and small (facies F) varieties. The large stromatolite columns in the vertical section often bear a vertical crack system irrespective of orientation. In bedding-parallel section, traces of three such vertical cracks, about 120°

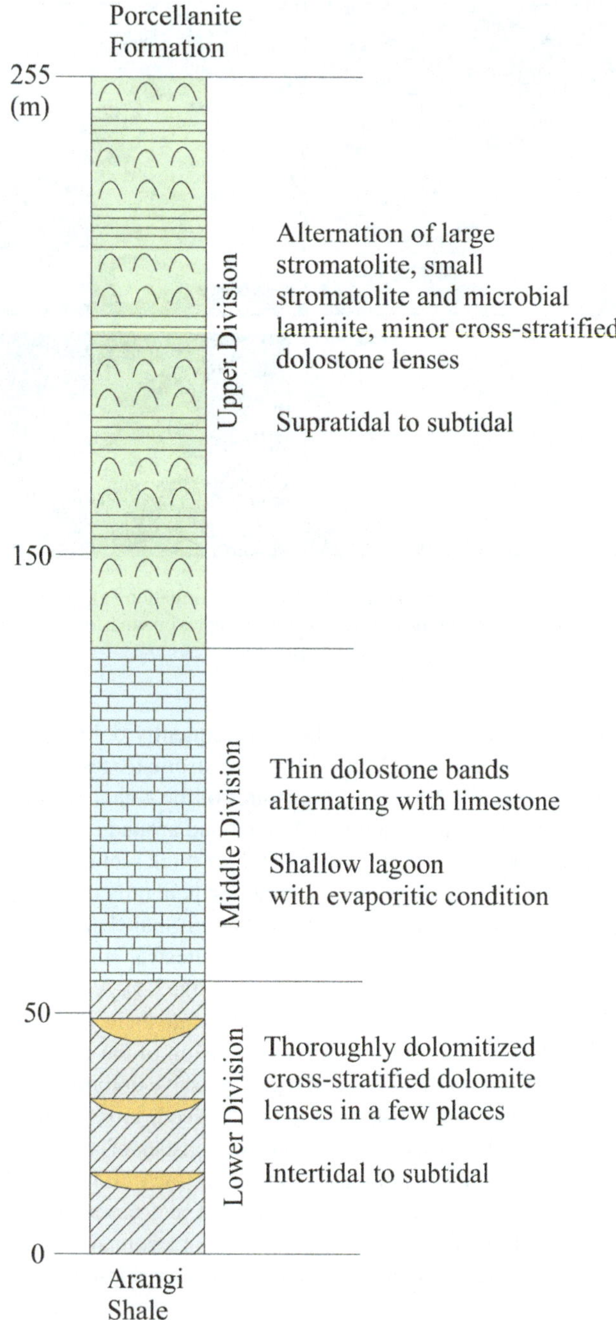

Fig. 2.4 Facies disposition within three broad subdivisions of the Kajrahat Limestone

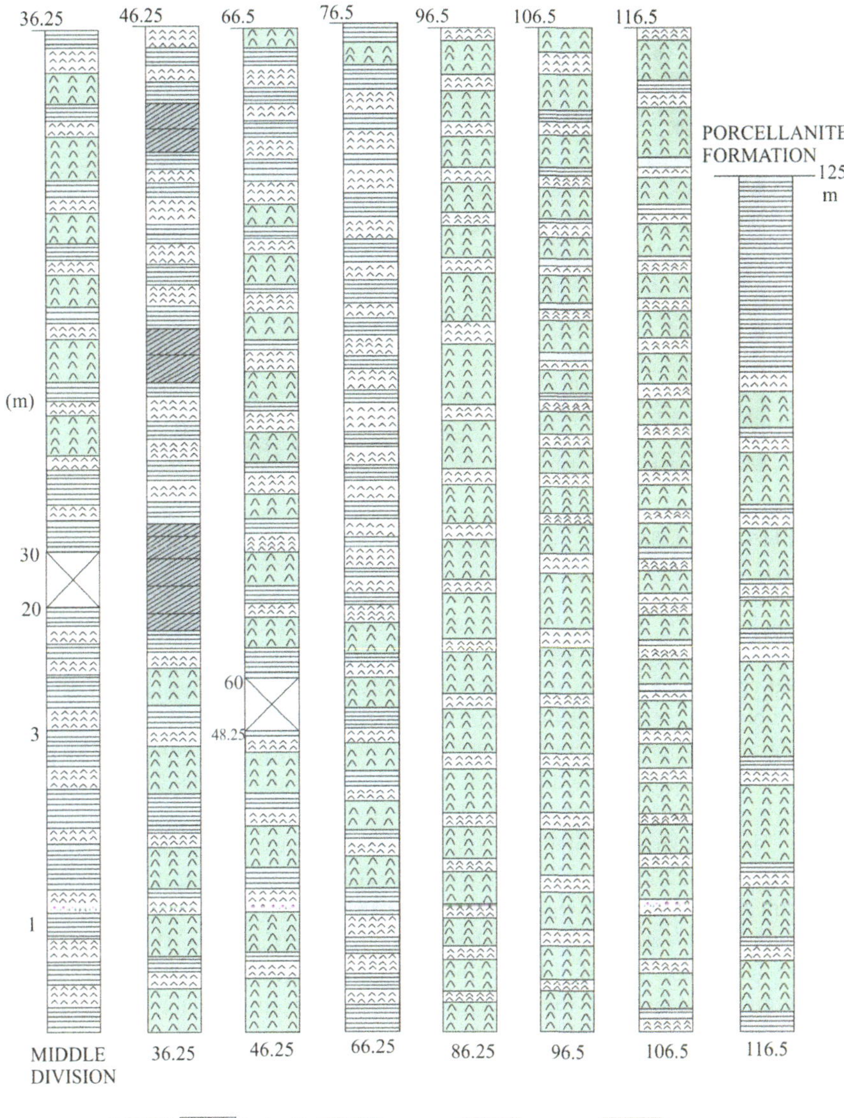

Fig. 2.5 Vertical section showing details of facies succession in the upper division of the Kajrahat Limestone. The section is bound by the middle division of the Kajrahat Limestone below and the Porcellanite Formation above. Note the cyclicity defined by large and small stromatolites and microbial laminate

Fig. 2.6 Large-scale stromatolitic columns with intercolumnar area within the Kajrahat Limestone (hammer handle for scale)

apart, can be seen (Fig. 2.6). The smaller stromatolites are mostly inclined and generally branching in nature (facies F, Table 2.2). Crinkled microbial laminae, bearing V-shaped cracks (facies G), occur above small stromatolites. The facies becomes considerably thicker towards the top of the upper division (Fig. 2.5).

The large-scale cross-stratified dolomite bodies occur as channel fills. The presence of cross-stratified dolomite beds in both lower and upper divisions of the Kajrahat Limestone indicates a shallow marine environment of deposition. The occurrence of oriented stromatolites corroborates this interpretation. Evaporitic gypsum pseudomorphs in dolomite beds in facies D and abundant shrinkage cracks in microbial laminae in facies G indicate periods of extreme shallowing. The facies G with microbial laminite represents the shallowest condition amongst all. The conical stromatolites (facies E), devoid of branching or preferred orientation, represents a subtidal setting (Grotzinger 1986; Southgate 1989; Altermann and Herbig 1991). The transition from large stromatolites (facies E) to small stromatolites (facies F) is gradational as stromatolites progressively decrease in size with gradual shallowing (Banerjee et al. 2007). The reduction of stromatolite height relates to a decrease in water depth (Beukes and Lowe 1989; Sarkar and Bose 1992; Glumac and Walker 1997). The microbial laminite with shrinkage cracks is characteristic of supratidal deposits (Altermann and Herbig 1991). As this facies thickens abnormally towards the top of the upper division, the Kajrahat Limestone represents an overall shallowing-upward succession.

Table 2.2 Description and interpretation of facies constituting the Kajrahat Limestone in Kuteswar–Dhanwahi area

Facies	Description	Interpretation
G	The facies dominantly consists of crinkly laminated limestones; bears V-shaped desiccation cracks; usually occurs at the top of small stromatolite facies. Towards the top of the stromatolite succession, the facies becomes abnormally thicker. Otherwise, thickness of the facies varies from 5 to 40 cm. It is dominantly constituted by calcimicrites	Resembles Microbial laminite. Desiccation cracks indicate emergence. Likely to be deposited in a supratidal setting
F	It consists of small stromatolites with average height 3–5 cm and head diameter 2 cm. They usually occur above the large stromatolites and underlies facies G. It bears axial desiccation cracks. Generally branching in nature and are inclined. Stromatolite inclination suggests northwesterly paleocurrent direction. Average thickness of the smaller stromatolite facies is 16 cm. It is constituted by calcimicrites	Desiccation cracks indicate emergence. It represents intertidal deposits
E	It consists of large stromatolites with average height and head diameter 20 cm and 7 cm, respectively; generally conical in nature, devoid of branching and inclination; may have axial desiccation cracks. Intercolumnar areas are narrow and filled by small pebble to mud size stromatolite fragments that are selectively dolomitized. Average thickness for the large stromatolite is 38 cm. It is constituted by calcimicrites	Lower intertidal to subtidal deposits. Greater height of stromatolite columns suggests deeper bathymetry
D	It consists of yellowish-grey-coloured dolostone; alternates with facies C. They usually have sheet-like geometry and sharp base. Average bed thickness is ~4 cm. It is comprised of subhedral to euhedral non-ferroan dolospars. Pseudomorphs of gypsum rosettes are abundant. Typical swallow-tale structures are locally recognizable. Despite the general massiveness, thin laminae can be seen under microscope and rosette structures are commonly upright on these planes	Intertidal to shallow lagoonal depositional conditions, frequent evaporitic condition prevailed
C	Consists of dark-grey-coloured, faintly laminated limestone. Thickness of facies decreases upwards from 1.3 m to 25 cm. Comprised of dolomicrites/micritic limestone, locally recrystallised	Intertidal to shallow lagoonal depositional conditions, frequently evaporitic

(continued)

Table 2.2 (continued)

Facies	Description	Interpretation
B	The facies consists of isolated yellowish-buff-coloured dolostone bodies. The dolostone beds are usually cross-stratified. Average cross-set thickness of 35 cm. Cross-stratifications indicate northwesterly paleocurrent direction. Consists of subhedral non-ferroan dolospars	Shallow marine carbonate sands (intertidal-subtidal)
A	The facies consists of yellowish-buff-coloured massive dolostone. Sedimentary structures are poorly preserved. It is made up of subhedral non-ferroan dolospars	The dolostone is thoroughly recrystallized. The rarity of current structures indicates a low-energy depositional environment

2.4 Porcellanite Formation

The Porcellanite Formation is up to 400 m thick, consisting of volcaniclastic tuff, pyroclastic flow and surge deposits (Banerjee 1997; Roy and Banerjee 2002). While Srivastava (1977) recognized subaerial deposition, Banerjee (1997) recorded frequent subaqueous to subaerial transitions within the Porcellanite Formation (Figs. 2.7, 2.8). The Porcellanite Formation is considered predominantly as felsic volcaniclastics by early workers (Auden 1933; Ghosh 1971). Mishra et al. (2018) considered Plinian-type eruptions from isolated vents causing widespread deposition of volcanic tuffs (see also Basu and Bickford 2015; Bickford et al. 2017). It is well exposed south of Chorhat and also around Chopan (Fig. 1.1). The lower contact of the Porcellanite Formation is poorly exposed. But the immediately underlying Kajrahat Limestone in its topmost part exhibits inclined small stromatolites and microbial laminites with abundant desiccation cracks. The Porcellanite Formation consists of the following facies around Chopan area (Table 2.3, Figs. 2.7, 2.8; for details, see Roy and Banerjee 2002).

The facies succession reveals recurrent vertical transitions between subaqueous and subaerial facies. Water depth possibly remained very shallow. The highly explosive nature of the volcanism is evident by the complete absence of purely magmatic rock of rhyolitic composition anywhere in the Son valley and abundant pumice shards in the constituent facies. The facies A with a dark and white band possibly represents alternating hot and cold ash deposition. These alternating welded and non-welded bands possibly indicate the intermittent, pulsating, Plinian-type of eruptions in subaerial conditions (Cas and Wright 1987; Schmincke and van den Boggard 1991; Orton 1996). Features indicative of heat retention, i.e. welded and flattened nature of the pumice shards in the black layers and columnar jointing, corroborate the subaerial origin for this facies. The close association of facies B with facies A and similar pumice concentration suggest possible subaerial deposition of the former. Columnar jointing and welded tuff are characteristics for both the facies. The presence of outsized bombs in some facies and heat retention features, i.e. welded nature

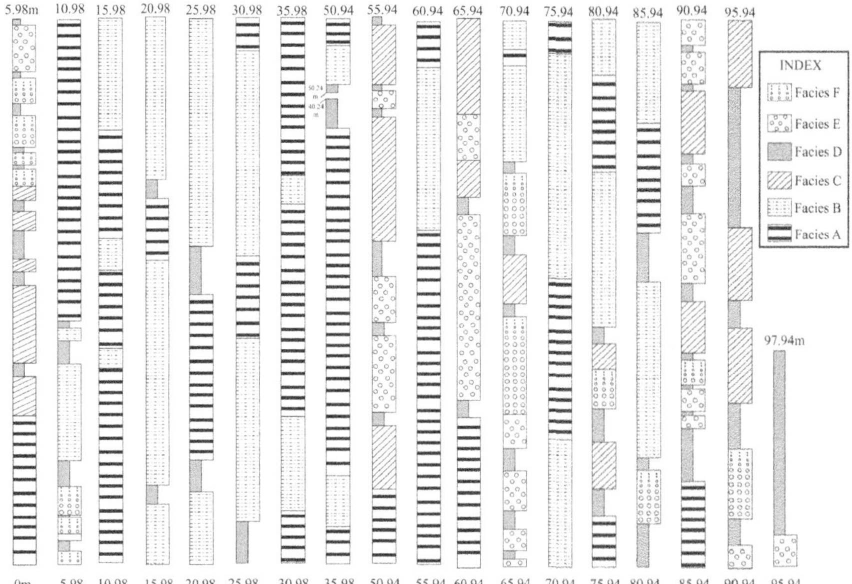

Fig. 2.7 Litholog showing vertical facies variation within the Porcellanite Formation in Chopan area (facies A—alternate black and white banded, facies B—crudely stratified, facies C—cross-stratified coarse-grained, facies D—shale, facies E—clast-rich, facies F—graded)

and columnar jointing, suggests proximity to volcanic vents. Remaining four facies are devoid of columnar jointing and welded tuff, suggesting subaqueous deposition. Facies C exhibits wave reworking of relatively coarse tuffs, suggesting deposition within the fair weather wave base. The deposition must have taken place in a shallow setting so that the impact of bombs could generate the ripples surrounding them. Facies D resembles normal mudrock and is inferred as subaqueous deposits marking breaks in the volcanic eruption. Facies E and F are related to mass flows of pyroclastic debris in a shallow, subaqueous condition. However, Chakraborty et al. (1996) considered a deep marine origin for the Porcellanite Formation based on the mass flow deposits. However, such deposits may generate in shallow marine depositional setting and even in subaerial conditions in volcanic environments (Lajoie and Stix 1992). The frequent subaerial to subaqueous transition in the Porcellanite Formation may either represent cyclic fluctuation in sea level or suggest fluctuations in volcaniclastic output.

2.5 Kheinjua Formation

The Kheinjua Formation begins with a ~12-m-thick dark shale of offshore origin. The Kheinjua Formation is well exposed around Chorhat–Shikarganj sector and Jadunath-pur–Bandu sector in the east (Fig. 2.9). The Koldaha Shale consists of sandstone

Fig. 2.8 Alternations of black and white banded tuffs (**a**), climbing ripple lamination (**b**) rectangular clasts (**c**) and columnar joints (**d**) within the Porcellanite Formation (coin diameter = 1.9 cm, matchstick length = 4.4 cm and hammer length = 38 cm)

and shale alternations. Both frequency and thickness of sandstone beds increase upwards, with local intraclastic breccia wedge (Bose et al. 1997, 2001; Banerjee 2000; Banerjee and Jeevankumar 2003, 2005; Sarkar et al. 1996; Samanta et al. 2016). The Koldaha Shale gradationally passes upwards into the Chorhat Sandstone. The latter comprises mainly amalgamated shallow marine storm beds (Seilacher et al. 1998; Sarkar et al. 1996, 2006).

2.5.1 Koldaha Shale

In its type area Koldaha, the shale consists of four facies, viz. shale, heterolithic, fine sandstone and coarse sandstone, recurring at two vertical segments (Table 2.4, Fig. 2.9a, b); For details, see Sarkar et al. 1996; Bose et al. 1997; Banerjee 2000). The shale facies contains thin (<2 cm thick) sandstone interbeds and gradationally passes over to the heterolithic facies. The heterolithic facies consists of alternations between shale and sandstone, but the sandstone beds are thicker (>3 cm and up to 38 cm) compared to those in the shale facies. Sandstone beds in both facies bear wave imprints (Fig. 2.10). Heterolithic and shale facies represent deposits on the shelf, proximal and distal, respectively. The heterolithic facies exhibits an overall

Table 2.3 Description and interpretation of facies constituting the Porcellanite Formation around Chopan area

Facies	Description	Interpretations
Graded tuff (facies F)	This facies is characterized by normally graded, light grey porcellanites. It occurs in association with the clast-rich porcellanite (facies G). Lower part of the beds may appear massive. Beds are broadly tabular in nature, thickness varying from 10 cm to 50 m. The facies is completely devoid of columnar joints and welded tuff	The absence of columnar joints and non-welded nature of the shards suggest subaqueous deposition. Their close association with facies E suggests common mode of origin for both the facies. Laminar debris flow possibly underwent body transformation by further intake of water and generated sediment gravity flows
Clast-rich tuff (facies E)	This facies is characterized by dark and light grey clasts set in a fine-grained, light-coloured matrix and appears as conglomerate. The clasts are clearly derived from facies A and B. The broadly rectangular, sharp-edged clasts range in size from 1 to 8 cm of length. The bed geometry is lenticular with sharp lower and upper contacts. Columnar joint is completely absent within the facies	The clasts being entirely derived from facies A and B, this facies is considered as epiclastic deposit which formed during inter-eruptive phases of volcanism. The sharp-edged and rectangular shape of the clasts possibly owes their origin to columnar jointing. The facies possibly represents pyroclastic gravity flow deposits. Deposition possibly took place from cold density currents similar to laminar debris flows
Shale (facies D)	Greyish to greenish, fissile shales alternate with the porcellanite bands at several levels. Predominantly fine-grained, fissile nature distinguishes the shales with other constituent facies of the Porcellanite Formation. No visible sedimentary structures other than faint laminations found in this facies	This facies is inferred to be deposited in a quiet water depositional setting by predominant suspension fallout of very fine sediments. This very fine-grained facies, occurring in association with cross-stratified porcellanite and clast-rich porcellanite represent deposition in a quiet water environment
Cross-stratified coarse-grained tuff (facies C)	The facies is characterized by relatively coarse-grained, chevron cross-stratified tuffs. It is commonly associated with the shale facies. Bed geometry is broadly tabular. Lower part of the beds may appear massive. Load casts may occur at the sole of the porcellanite beds	Presence of chevron cross-stratifications definitely suggests wave reworking and to-and-fro particle movement. Cold state of deposition is indicated by lack of welding and columnar joints within the facies. Massive appearance near the basal part and presence of load casts at the base suggest rapid deposition of pyroclasts

(continued)

Table 2.3 (continued)

Facies	Description	Interpretations
Crudely stratified tuff (facies B)	This is characterized by dark-coloured tuff almost similar to facies A, but internally displaying crude laminae. The facies also exhibits columnar jointing and welding. Both lower and upper contacts of the beds are non-erosional. Layer confined micro-faults can be recognized	Close association of this facies with facies A suggests a common mode of origin for both. The absence of white bands unlike facies A suggests steady and continuous eruption. Slight fluctuations in the discharge rate might be the cause of the crude laminae observed within the facies
Alternating black and white banded tuff (facies A)	Laterally persistent alternate bands of light and dark tuffaceous materials are conspicuous within the facies. Bombs of 6–15 cm are present on the bedding planes within the dark layer. The contacts between black and white bands are sharp. Columnar jointing can be observed within this facies. Asymmetric eolian ripples may occur at the top of white bands in places	Alternate persistent banding in centimetre scale reflects products of pyroclastic fall type of deposits. Oversized bombs of various sizes suggest possible near vent deposits. The alternate welded (dark) and non-welded (white layers) indicates the deposition of hot and cold pyroclasts and pulsating volcanic eruption

coarsening-upward trend and gradationally passes upwards to the fine sandstone facies. The latter consists of coarsening-upward, thoroughly wave-imprinted sandstone and is almost devoid of mud. The fine sandstone facies possibly represents the upper shoreface setting. The coarse sandstone facies, at the top, is mineralogically and texturally immature, representing a fluvial deposit. The overall gradational trend without any incision at the top of the coarse sandstone facies suggests a braid plain origin for this facies (cf. Bose et al. 2001; Banerjee and Jeevankumar 2005). The coarse sandstone facies at the mid-level of the Koldaha Shale is again sharply overlain by the shale facies, with granular lag at their contact (Fig. 2.9a, b). Facies stacking reveals the existence of two decametre-scale, overall coarsening-upward segments within the Kheinjua Formation (Fig. 2.9a, b). These segments are separated from each other by the thin and sheet-like granular lag. As the granule bed immediately passes upwards to a shale facies of distal shelf origin, it is considered as a transgressive lag.

2.5.2 *Chorhat Sandstone*

The Chorhat Sandstone gradationally overlies the Koldaha Shale and it is best exposed on the side of the metalled road between Chorhat and Rampur. The Chorhat Sandstone consists of three nonrecurring facies in the type area near Chorhat

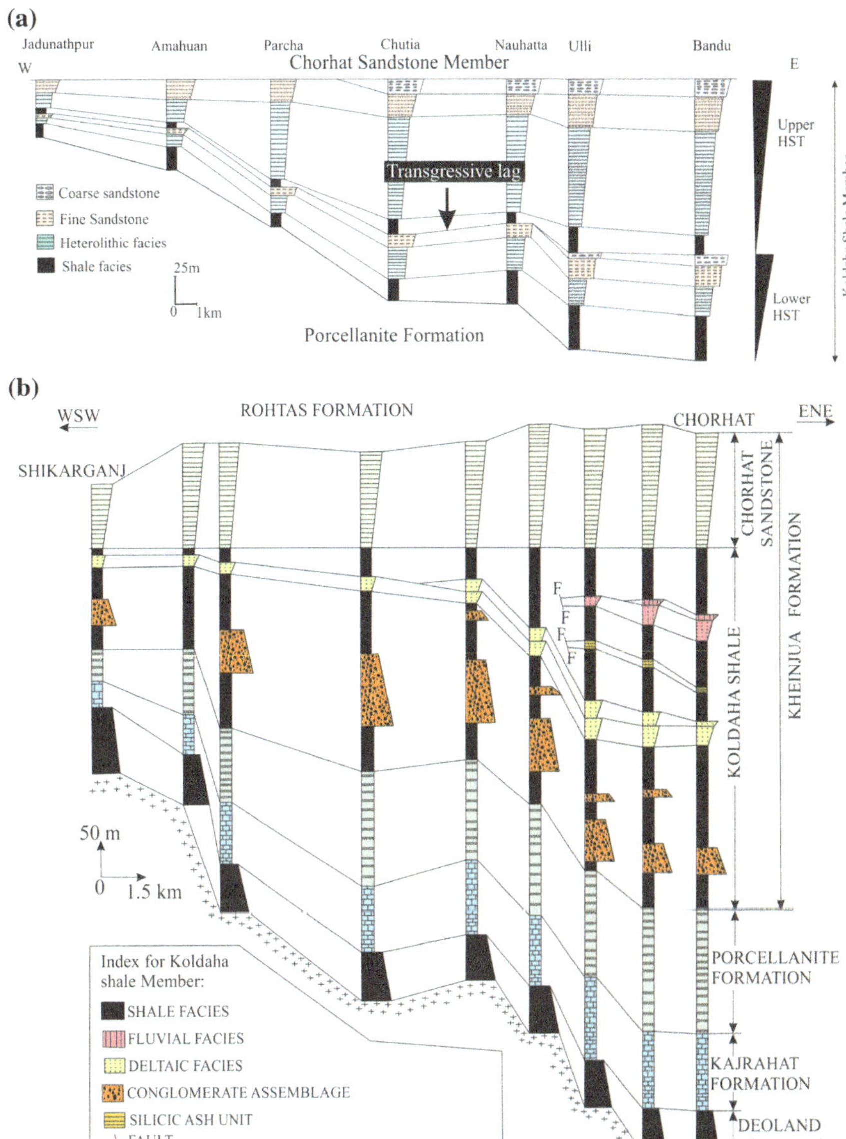

Fig. 2.9 Panel diagram showing lateral facies variation within the Koldaha Shale in the eastern part of the Son Valley (location of Nauhatta is indicated in Fig. 1.1). Note two HSTs separated by a granular lag (**a**), Panel diagram showing vertical variation of sediment thickness within the Lower Vindhyan between Chorhat and Shikarganj (**b**) (note westward thinning of the constituent facies of the Kheinjua Formation. For locations, see Fig. 3.1

Table 2.4 Description and interpretation of facies constituting the Koldaha Shale around Chorhat area

Facies	Description	Interpretations
Coarse sandstone	It is texturally and mineralogically immature granular sandstone, bears tabular and trough cross-strata; cross-strata show unimodal orientation. Generally, it is arranged in fining-upward channel-filled bodies bounded between successive broadly undulated master erosion surfaces	Fluvial braid plain deposit with negligible lithological difference between channel and interchannel areas
Fine sandstone	Moderately well-sorted fine sandstone characterized by planar and wavy laminae; locally tabular cross-bedding; Facies units are coarsening upwards with thin (~1.5 cm) shale interbeds at base. Their tops bear slump folds and convolutes	Predominantly shoreface depositional environment, extending up to the relatively steep upper shoreface
Heterolithic	Rhythmic interbedding between shale and fine-grained sandstone (3–38 cm thick) having sharp base sculpted with prominent tool marks (bipolar prod marks), planar and wavy laminae inside. Wave ripples occur on top of sandstone beds. Stromatolites and microbial-laminated carbonate bodies occur locally. The facies units are coarsening upwards. Sandstone beds thicken upwards at the expense of shale interbeds	Inner shelf, mostly between fair weather and storm wave bases
Shale	Greenish-grey shale with rare thin (2 cm) planar-laminated fine sandstone beds having fine tool-marked sharp planar base. Top of beds undulated with small wave ripples	Outer shelf, near the storm wave base

(Table 2.5; Fig. 2.11). It represents predominantly wave-dominated marine shelf deposit (Sarkar et al. 2006), bordered by a coastal flat with eolian sand sheets. The lowermost facies (Facies A) is characterized by vertical stacking of 35–40-cm-thick beds of light coloured, overall graded, tabular sandstone, alternating with siltstone beds (<5 cm thick). The beds have sharp and erosional bases, riddled with gutters and prod marks. The sandstone beds of this facies are characterized by hummocky cross-stratification, quasi-planar strata and wave ripples (Fig. 2.12). The middle one (Facies B) is characterized by well-sorted, fine-grained, wave-rippled sandstone with

Fig. 2.10 Characteristics features of the Koldaha Shale: Shale with thin sandstone interbeds (**a**), shale and sandstone alternation (**b**), wavy laminae within a sandstone bed (**c**), coarse sandstone alternating with siltstone beds (**d**) and planar-laminated fine sandstone (**e**) (marker length in c = 13.5 cm, pen length in c = 14.5 cm, outcrop width in d = 40 cm)

emergence features towards the top (Fig. 2.12). The sandstones of this facies like those of Facies A are well sorted and are completely devoid of mud. The uppermost unit (Facies C) occurs locally and consists of well-sorted sandstone with various eolian features.

2.6 Rohtas Formation

The Chorhat Sandstone is overlain sharply by the offshore-originated Rampur Shale, which makes up the lower part of the Rohtas Formation (Sarkar et al. 2002a; Banerjee et al. 2006a). The Rampur Shale becomes richer in organic content as it fines upwards, and incorporates tuffaceous deposits towards the top (Sarkar et al. 2002a; Banerjee et al. 2006a; Sur et al. 2006). Good exposures of the Rampur Shale around Kudari area allow detailed facies analysis (Table 2.6, Figs. 2.13, 2.14; for details, see Sarkar

Table 2.5 Description and interpretation of facies constituting the Chorhat Sandstone around Chorhat area

Facies	Description	Interpretation
C	It consists of medium-grained sandstone characterized mainly by adhesion laminae and translatent strata. Isolated cross-sets (<32 cm thick) with inversely graded grainflow laminae alternating with finer grained grainfall laminae locally present. Maximum thickness of facies is ~8 m, and it occurs locally	Eolian sand sheet formed presumably in an erg-margin wet system, perhaps along a coastal fringe (supratidal erg-margin)
B	Fine- to medium-grained sandstone beds generally thicker than 10 cm, often amalgamated, massive or quasi-planar and wavy-laminated, the latter intervened by wavy erosion surfaces. Wave ripples often preserved on bed top. The sandstone bed surfaces at the top ~10 m of the facies unit bear ripples migrating along troughs of larger ripples, divergent parting lineation, and spectacularly preserved rill marks. Muddy siltstone, generally thinner than a centimetre, locally occurs within this part. Reddish mud clasts concentrate at the base of many sandstone beds. Wart marks and adhesion ripples occur locally. Maximum thickness of the facies unit is 45 m	Deposition presumably took place on shallow shelf, mostly within fair weather wave base. Towards top depositional surface got exposed intermittently (subtidal to intertidal)
A	Characterized by fine-grained sandstone–siltstone interbeds. Sandstone beds are tabular in shape, often less than a centimetre thick; relatively thicker beds are overall graded, have sharp erosional bases riddled with gutters and prods followed up by quasi-planar laminae, hummocky cross-stratification and finally wave ripples on bed tops. Amalgamation between sandstone beds is not uncommon. The facies thins and eventually disappears west of Chorhat. Maximum facies unit thickness is 15 m	Siltstone beds are autochthonous, while sandstone beds formed during high energy events, probably storms. Common thinness of storm beds suggests deposition near the storm wave base (deeper subtidal)

et al. 2002a; for locations, see Fig. 3.1). The top part of the Rampur Shale, consisting of alternations between organic-rich shale and thin limestone beds, is well exposed around Rampur area. The Rampur Shale gradationally passes over to the Rohtas Limestone. The distinctly fining-upward marine Rampur Shale, lacking emergence features, is overall transgressive in nature. The organic-rich and pyritiferous shale, containing pyroclastics on top, represent a condensed interval (cf. Kidwell 1991; Bose et al. 2001).

Exposures of the Rohtas Limestone are fairly continuous, starting from Amjhore in the east to Katni in the west of the Son valley area, through Chopan and Maihar (Fig. 1.1). Workable exposures within limestone quarries allow detailed facies analysis. The facies assemblage of the Rohtas Limestone in U.P. and Bihar area is broadly similar to that of the western sector. A brief description of the constituent facies of the Rohtas limestone is provided in Table 2.7 (for details, see Banerjee et al. 2005, 2006b; Banerjee and Jeevankumar 2007). Black shale facies, crinkly laminated limestone facies and nodular limestone facies comprise the lower part of the Rohtas Limestone

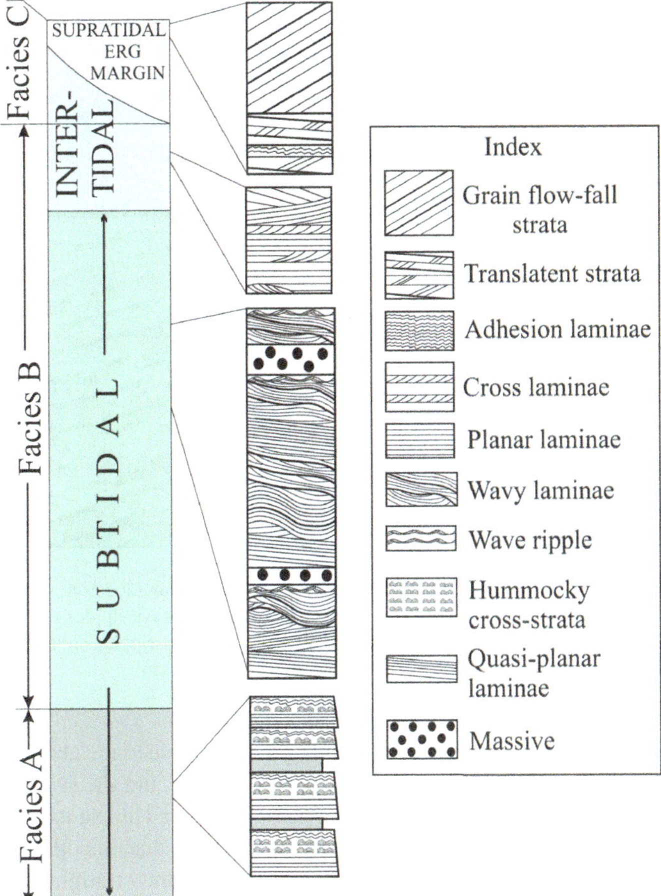

Fig. 2.11 Stratigraphic frame of the Chorhat Sandstone differentiated in facies superposed one over the other. Note structural assemblages for different facies

(Figs. 2.15, 2.16). The crinkly laminated limestone facies alternates with the black shale facies. The absence of desiccation features in both these facies indicates their outer shelf origin. The nodular limestone facies occurs mostly in association with the black shale facies and represents the early diagenetic modification of the microbial laminite.

The top segment of Rohtas Limestone consisting of thinly laminated heterolithic facies, plane-laminated limestone facies and wavy-laminated limestone facies alternating with the grey shale facies, possibly represents the shallower shelf above the storm wave base. The thinly laminated heterolithic facies, showing millimetre-scale alternations between black shale and sheet limestone, remains confined to the intermediate level mostly in association with grey shale. The high TOC content and pyrite framboids within the black shale laminae in this facies indicate temporary develop-

Fig. 2.12 Characteristic features of the Chorhat Sandstone: Tabular sandstone alternating with siltstone beds (**a**) vertical section showing quasi-planar strata (**b**), wave ripples on the surface of a sandstone bed showing tuning-fork bifurcations (**c**) and secondary ripples along the troughs of primary forms (pen length in b = 14.5 cm, coin diameter in c = 2.6 cm)

ment of anoxic condition at a moderate depth, possibly close to the chemocline. The other two shallow facies, viz. plane-laminated limestone and the wavy-laminated limestone, are inferred as allochthonous. The plane-laminated limestone facies shows a distinct preference for relatively lower stratigraphic level than the other facies. The wavy-laminated limestone facies generally thickens up the stratigraphic column. Relatively greater depth facilitates the deposition of the plane-laminated limestone facies and a shallower depth (possibly lower shoreface) allows the formation of the wavy-laminated limestone facies. Nonetheless, sediment appears to be delivered by the same agent, possibly storm waves, in case of both the facies.

The Rohtas Limestone in the eastern sector of the Son Valley is a shallowing-upward succession developed on an epeiric shelf. It rests on a condensed zone deposit and is truncated at top by a basin-wide unconformity. Bounded by a condensed zone below and an unconformity above, the Rohtas Limestone shows overall prograding nature in a truly basinal scale and represents a highstand systems tract.

2.7   Kaimur Formation

The Upper Vindhyan sequence comprises three formations, viz. Kaimur, Rewa and Bhander in ascending order of succession. The siliciclastic Kaimur Formation is

Table 2.6 Description and interpretation of facies constituting the Rampur Shale around Kudari area

Facies	Description	Interpretation
Facies C	The shale is distinctly black in colour and fissile in nature, but generally lacks siltstone inter-laminae. Its organic carbon content locally exceeds 3%. The shale contains many darker blebs and is rich in pyrite. Felsic pyroclastic deposits of about 6 m thick occur at two levels	The black shale facies without any sand/silt stringers possibly deposited on a distal part of the shelf
Facies B	It consists of greenish-grey shale with isolated quartz-rich and fine-grained sandstones. It is about 60 m thick. The shale is olive green in colour and contains about 1.1% organic carbon. Millimetre-scale, siltstone laminae within shales may exhibit either massive or faintly planar laminae. The sandstone occurs largely within small gutter casts. The sandstone lenses exhibit planar laminae, ripple cross-laminae and wavy laminae. These beds also show overall grading with mud clast concentrations at their bases. Wave ripples, with trends similar to those of the underlying facies, occur locally on bed tops. Silt-filled gutters are thinner but wider (up to 1 m) towards the top of the facies unit. Wave ripples and convolute laminae occur locally. Sandstone/siltstone gutter fills progressively decrease in thickness upwards and eventually disappear, giving way to a 17-m-thick black shale (facies C)	The shale indicates suspension fallout deposit on shelf, while the sandstone and siltstone gutter casts formed during high energy events such as storms. Decrease in width/depth ratios of gutters upwards indicates gradual weakening of high energy flows. The facies presumably developed in a mid-shelf area, within reach of storm wave base
Facies A	The basal facies consists of buff-coloured, quartz-rich, fine-grained sandstone that is faintly cross-stratified. It is about 8 m in thickness. The cross-stratification (<14 cm set thickness) is planar, quasi-planar and rarely hummocky. Wave ripples occur at the top of the sandstone beds. Wave ripple crests are generally aligned in northwest–southeast direction. Laterally discontinuous, cm-thick silty shales alternate with sandstones. Wrinkle marks are common locally on bedding surfaces	Deposition possibly took place in lower shoreface/inner shelf transition environment. Quasi-planar stratification and hummocky cross-stratification indicates storm influence (Arnott 1993). The subordinate mud may be deposited as suspension fallout during waning stages of storms and in between storms

the lowermost stratigraphic unit of the Upper Vindhyan Group. The Kaimur Formation is divisible into two parts, viz. upper and lower (Fig. 2.17). The lower part of the Kaimur Formation consists of Lower Quartzite (also known as Sasaram Sandstone Member) and Bijaigarh Shale members, while Scarp Sandstone and Dhandraul Quartzite members are present at the upper part (Fig. 2.18). The Lower Quartzite maintains an unconformable relationship with the underlying Rohtas Limestone of Semri Group. A silicified shale and a sandstone horizon (known as Upper Quartzite)

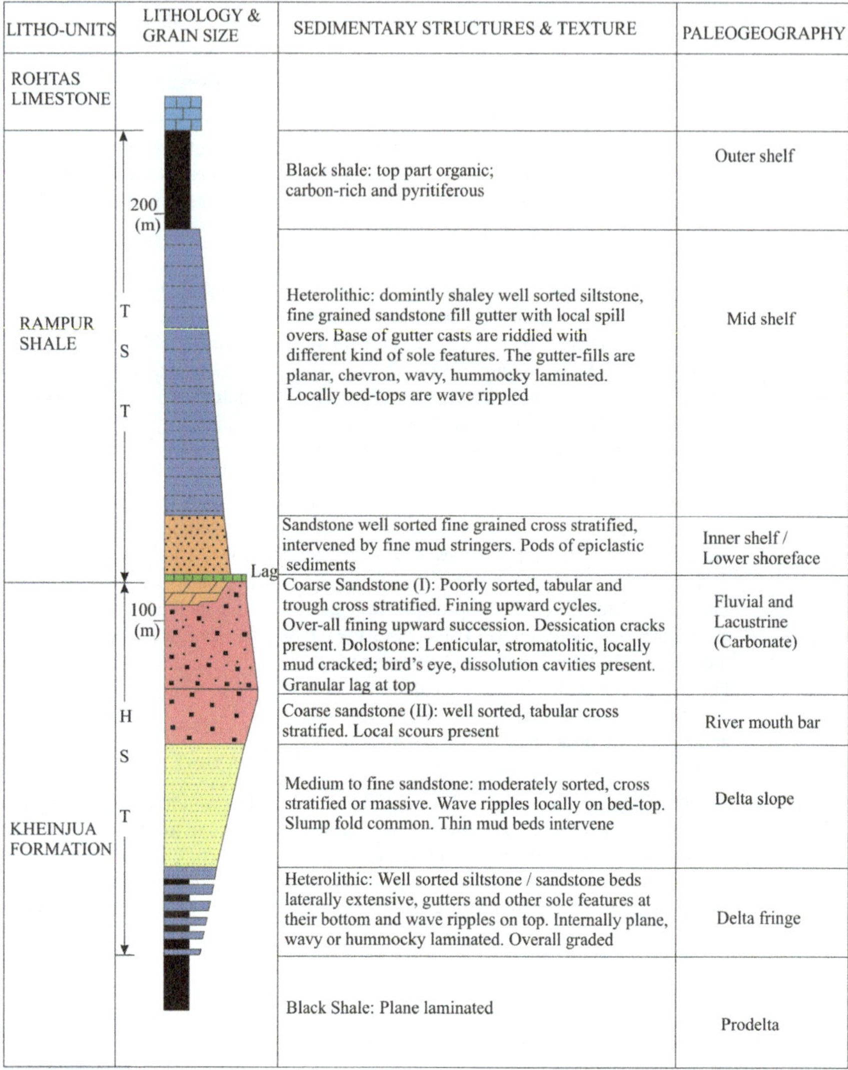

LITHO-UNITS	LITHOLOGY & GRAIN SIZE	SEDIMENTARY STRUCTURES & TEXTURE	PALEOGEOGRAPHY
ROHTAS LIMESTONE			
RAMPUR SHALE		Black shale: top part organic; carbon-rich and pyritiferous	Outer shelf
		Heterolithic: domintly shaley well sorted siltstone, fine grained sandstone fill gutter with local spill overs. Base of gutter casts are riddled with different kind of sole features. The gutter-fills are planar, chevron, wavy, hummocky laminated. Locally bed-tops are wave rippled	Mid shelf
		Sandstone well sorted fine grained cross stratified, intervened by fine mud stringers. Pods of epiclastic sediments	Inner shelf / Lower shoreface
KHEINJUA FORMATION		Coarse Sandstone (I): Poorly sorted, tabular and trough cross stratified. Fining upward cycles. Over-all fining upward succession. Dessication cracks present. Dolostone: Lenticular, stromatolitic, locally mud cracked; bird's eye, dissolution cavities present. Granular lag at top	Fluvial and Lacustrine (Carbonate)
		Coarse sandstone (II): well sorted, tabular cross stratified. Local scours present	River mouth bar
		Medium to fine sandstone: moderately sorted, cross stratified or massive. Wave ripples locally on bed-top. Slump fold common. Thin mud beds intervene	Delta slope
		Heterolithic: Well sorted siltstone / sandstone beds laterally extensive, gutters and other sole features at their bottom and wave ripples on top. Internally plane, wavy or hummocky laminated. Overall graded	Delta fringe
		Black Shale: Plane laminated	Prodelta

Fig. 2.13 Vertical facies sequence in the Rampur Shale and its immediately underlying formation, depicting variations in lithology, average relative grain size, structures, textures and inferred paleogeography

occurs locally (cf. Chakraborty 1994). The Kaimur Formation is well exposed in Churk and Mirzapur area in U.P. and Mohonia (near Chorhat) in M.P.

The lower part of the Kaimur Formation including the Lower Quartzite Member and the Bijaigarh Shale Member is well exposed in the southeastern sector of the Vindhyan outcrop (Figs. 2.18, 2.19, 2.20). Facies constituting these two members are provided in Table 2.8. The Lower Quartzite Member exhibits a fining-upward trend. The sandstone of tidal bar origin occurs at the bottom part of the formation

Fig. 2.14 Photographs showing amalgamated sandstone (**a**) cross-stratified sandstone (**b**), wave-rippled sandstone (**c**), vertical section showing a gutter cast (**d**), sole marks below a gutter cast (**e**), siltstone beds within shale (**f**), green shale alternating with sandstone (**g**), wave ripples on the bed surface of a sandstone (**h**), alternation between black shale and pyroclastic beds (**i**) alternation between black shale and limestone (**j**) within Rampur Shale (pen length in a, d, g, i, j = 14.5 cm, Swiss knife length in b, e, h = 9.1 cm)

Table 2.7 Description and interpretation of facies constituting the Rohtas Limestone around Ghurma area

Facies	Description	Interpretation
Wavy-laminated limestone	These calcarenite beds exhibit three-dimensional hummocky cross-strata and ripple cross-laminae. However, the lower part of the beds may be massive, graded or parallel-laminated and contains conspicuous concentration of clasts (size 3–5 mm) at base. Occasionally, the sole of the beds is riddled with gutter casts. The bed tops are often wave-rippled; the ripple cross-laminae under them show subcritical to supercritical climb. Draping and offshoot laminae occur locally within the ripple lamina set. Ripples are asymmetric in vertical profile with straight crests and tuning-fork bifurcations	Sharp and erosional lower contacts of the beds, presence of sole features and basal clast concentration clearly indicate allochthonous origin of this facies. The massiveness and grading in the lower part of the facies suggest rapid suspension fallout. Unidirectional ripple cross-laminae indicate the presence of a current component in the flow. The presence of three-dimensional hummocks and small-scale features, like draping and offshoot laminae, clearly testify the oscillatory component of the flow dominating over the current component
Plane-laminated limestone	This facies consists of calcimicritic limestone beds of 30 cm to <1 m in thickness. The beds generally show tabular geometry and show planar laminae. However, thicker beds (>40 cm) may bear minute clasts and display massiveness or normal grading at their base; towards top part, the beds are planar-laminated	Deposited by a turbidite-like underflow during high energy flux. The underflow could have been generated by storm wave. Planar laminae can be generated by high flow regime traction current as well as by wave
Thinly laminated heterolithic facies	This facies consists of millimetre to centimetre-scale close alternations between black shale and micritic limestone laminae, laterally persistent and sheet-like in geometry. The black shale laminae are essentially comparable to the black shale facies having similar high TOC and framboidal pyrites	Substantial fluctuations in availability of dissolved oxygen appear to be the characteristic of the depositional environment. Possibly deposition took place at the level of chemocline in a stratified seawater column
Grey shale facies	It consists of grey-coloured shale, up to 50 cm in thickness. It occurs preferably at the upper part of the succession. Internally, it exhibits planar laminae	Absence of wave features and laterally persistent planar-laminated shale beds suggest slow deposition in a quiet depositional setting

(continued)

Table 2.7 (continued)

Facies	Description	Interpretation
Nodular limestone facies	This facies consists of decimeters to about a metre-thick, tabular limestone beds characterized by nodularity. Most of the nodules appear to be flattened, ellipsoidal in nature, elongated and are bed parallel. Some nodules have irregular shapes, while a few rare ones are, contrastingly, perfect spheres	Nodules are in situ in origin and are not transported from outside. The nodules are products of early diagenesis. Diagenetic overprint has rendered the primary character of the limestone
Crinkle-laminated limestone facies	It consists of crinkly laminated, argillaceous limestone. The beds are grey to greyish-buff-coloured, tabular in geometry, and are always bounded below and above by the black shale facies	Crinkled laminae represents microbial mat. Very fine grain size of the sediments indicates lack of agitation
Black shale facies	It consists of black-coloured, hard and compact shale that appears massive in the field. Thickness of the black shale facies units bounded below and above by other facies varies from <1 cm to >1 m. Total organic carbon (TOC) content averages 1.2% for the carbonaceous shale. Wavy–crinkly fabric is conspicuous within the carbonaceous laminae	The wavy–crinkly, pyritic and carbonaceous laminae are characteristic for the microbially originated black shales

(Fig. 2.19). This is gradationally followed upwards by a fining-upward shale–sandstone alternation. The shale becomes sand-free at the top, and organic carbon-rich, containing pyrite and locally phosphate (Fig. 2.20; Chakraborty 1995; Banerjee et al. 2006a; Sur et al. 2006; Schieber et al. 2007). The pyritiferous, organic carbon-rich shale is best exposed around Amjhore Mines (Fig. 1.1). A basin-wide reworked volcanic tuff/shale layers of ~10 m thick demarcates the bottom of the Upper Kaimur, which is also known as Bhagwar Shale.

The upper part of the Kaimur Formation including the Scarp Sandstone and Dhandraul Quartzite is well exposed in Kalinjar, Chitrakut, Mirzapur, Sasaram, Hanumana, Mohania, Govindgarh and Maihar (Fig. 2.21). The Scarp Sandstone is predominantly storm-influenced (Table 2.9). The Dhandraul Sandstone is widespread, coarsening-upward sandstone body covering the entire Son Valley and show extreme variation in thickness from 8 to 228 m (Fig. 2.18; Bhattacharyya and Morad 1993). Considering the lithology, sedimentary structures and bed geometry of the constituting facies, this member can be grouped into two facies assemblages (Table 2.10).

Earlier studies reveal an upward transition from a marine to a terrestrial regime within the Kaimur Formation (cf. Chakraborty and Bose 1992; Chakraborty 1993;

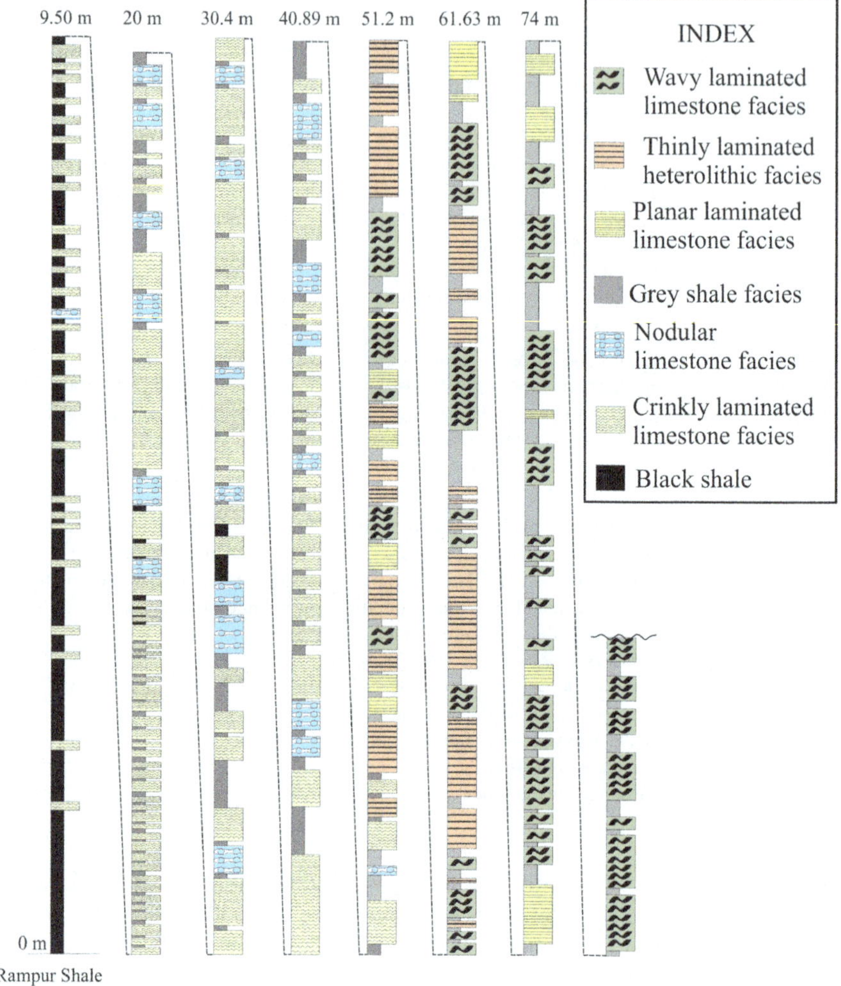

Fig. 2.15 Vertical log showing constituent facies of the Rohtas Limestone in the Ghurma Limestone Mines. Note the dominance of black shale, crinkle-laminated limestone and nodular limestone facies in the lower part and grey shale, thinly plane-laminated heterolithic limestone and wavy-laminated limestone facies towards the top

Bhattacharyya and Morad 1993). Wave and tidal processes dominate marine deposits, while the terrestrial record is dominated by the fluvial system with sporadic eolian imprints at places (Chakraborty 1993). The lower part of the Kaimur succession is transgressive, while the upper part is regressive. A gradual paleogeographic shift from shallow subtidal to deeper shelf has been considered for the transgressive lower part. The upper part of the Kaimur Formation, recording a transition from the inner shelf to the terrestrial environment, characterizes the regressive part (Chakraborty 1993). The

Fig. 2.16 Characteristic features of the Rohtas Limestone in profile: crinkle-laminated limestone (**a**), nodules within a bed (**b**), grey shale alternating with limestone (**c**) thinly laminated heterolithic facies (**d**), planar lamination (**e**) hummocky cross-stratification (**f**), ripple lamination (**g**) and flat pebble conglomerate (**h**) (pen length in a, b, c, d, g, h = 14.5 cm, hammer length in e = 38 cm)

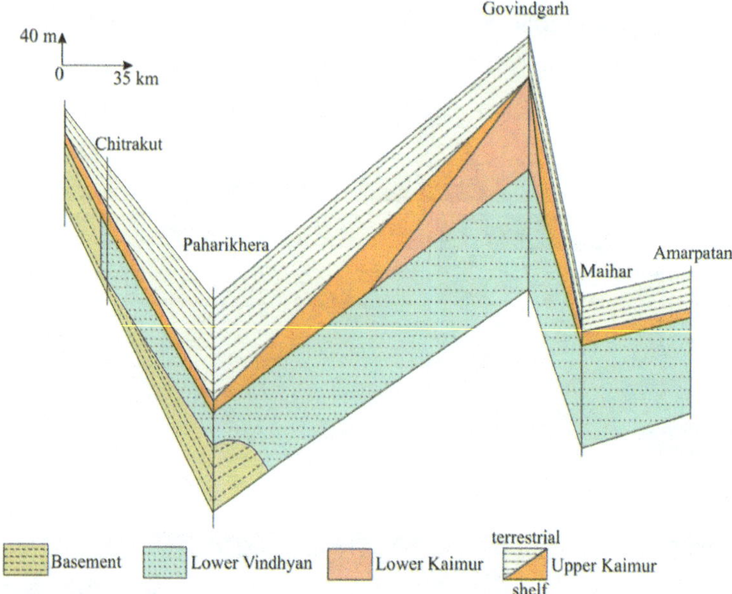

Fig. 2.17 Fence diagram showing lateral thickness variation within the Kaimur Formation, overlying the Lower Vindhyans and the basement rocks. Note onlapping of Kaimur shelf sediments on to the basement on the northern flank. Also note the distinct lenticularity of the Lower Kaimur

Upper Kaimur starts with a shelf storm succession (Chakraborty and Bose, 1992) that gradationally gives way upwards into a coarsening-upward succession consisting of fluvial and eolian deposits (Bhattacharyya and Morad 1993; Chakraborty 1993).

2.8　Rewa Formation

Rewa Formation overlies the Kaimur Formation. The Rewa Formation consists of sandstone and shale and is completely devoid of carbonate. The lower part of the formation is dominated by shale, while sandstone dominates the upper part (Fig. 2.22). Earlier the shale was divided into two members, viz. Panna Shale and Jhiri Shale, considering the presence of Lower Rewa Sandstone in between. However, the Lower Rewa Sandstone is a lensoid body and it does not occur in most places within the Son valley (Fig. 2.22; Chakraborty and Sarkar 2005). Later workers consider the Rewa Shale as a single unit (Chakraborty and Sarkar 2005). A thin granular lag occurs immediately above the Kaimur Formation, almost everywhere in the Son valley. The Rewa Shale overlies this granular lag (Fig. 2.23). A diamondiferous conglomerate occurs at the base of the Rewa Shale. Chakraborty et al. (1996) reported a volcaniclastic deposit above the conglomerate. The shelf succession of the Rewa Shale depicts stacking of shoaling-upward facies couplets separated from each other by

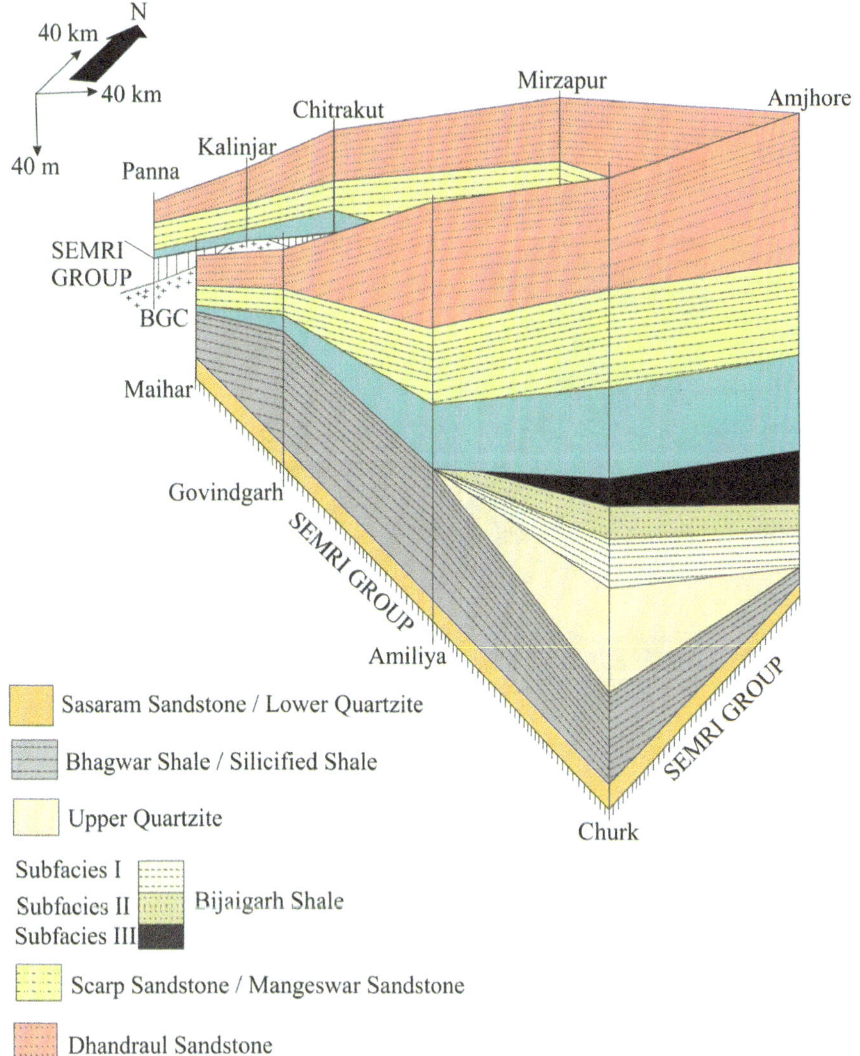

Fig. 2.18 Three-dimensional representation of different stratigraphic units of the Kaimur Formation. Note the pinching nature of the Bijaigarh Shale and Upper Quartzite

thin, well-sorted granular lags. The Rewa Shale passes upwards into the Rewa Sandstone, which has a coarsening-upward trend. Facies constituting the Rewa Shale is presented in Table 2.11. The Rewa Sandstone exhibits shallow marine-tidal deposits grading upwards into fluvial/eolian deposits (Bose and Chakraborty 1994).

The Upper Rewa Sandstone varies widely in thickness, maximum being at Drummondsganj in central India (Fig. 2.22). Exposure of both marine and non-marine part also occurs around Bhadanpur–Sarlanagar stretch, south of Maihar (Fig. 1.1).

Fig. 2.19 Sedimentological aspects of the Lower Quartzite: alternating thick–thin forests in tabular cross-stratified sandstone (**a**), herringbone cross-stratification (**b**), ripple lamination (**c**), planar-laminated sandstone (**d**), conglomerate (**e**) tabular cross-stratified sandstone (**f**), hummocky cross-stratified sandstone (**g**) (pen length in c, e = 14.5 cm, hammer length in f = 38 cm, width of outcrop in d = 60 cm)

In this stretch, all the facies associations constituting the Upper Rewa Sandstone are present. A detailed facies study around Bhadanpur–Sarlanagar stretch reveals an upward transition from marine to fluvial associations (Table 2.12; for details, see Bose and Chakraborty 1994). These two facies associations differ in stratal geometry, grain size, mineralogy and sorting. Paleocurrent pattern and direction also vary between the two facies associations. Overall, the Upper Rewa Sandstone consists of quartz arenite. Feldspar content is less than 5%. Mud is uncommon but mud

Fig. 2.20 Bijaigrah Shale in profile: black shale with thin lenses of siltstone (**a**), gutter cast within black shale (**b**) (pen length = 14 cm)

clasts are present within certain facies. The marginal marine association occurs at the lower part of the succession in Bhadanpur–Saralanagar area (Fig. 2.24). Sedimentary structures indicate actions of both tide and wave during the deposition of sediments. The marine association is dominated by fine-grained sediment showing rhythmic changes in thickness and style of cross-stratification. Features representing local current reversals suggest the action of tides. Beach deposits, on the other hand, occur locally and suggest wave actions. The major paleocurrent direction is consistently westwards. The fluvial association overlies the marginal marine association within this stretch (Fig. 2.25). The sandstone is coarse-grained, even granular within the fluvial association, containing occasional rip-up clasts. Sandstones of this association are strikingly less sorted compared to those of marginal marine facies association.

2.9 Bhander Formation

The Bhander Formation is the youngest member of the Upper Vindhyan Group, and it consists of five members, viz. Ganurgarh Shale, Bhander Limestone, Lower Bhander Sandstone, Sirbu Shale and Upper Bhander Sandstone. The Bhander Limestone is the only laterally persistent limestone horizon present within the Bhander Formation, and it is the only carbonate deposit in the entire Upper Vindhyan. It is bounded by the underlying Ganurgarh Shale and the overlying Lower Bhander Sandstone, both of which are the coastal deposit. The Bhander Limestone is a shallow shelf product with abundant stromatolites. The Lower Bhander Sandstone contains a high amount of mud and its exposure is patchy. The dominance of mudstone and the tan-coloured

Table 2.8 Description and interpretation of facies constituting the lower part of the Kaimur Formation

Facies	Description	Interpretation
Alternation of thick and thin tabular cross-stratified sandstone facies	Comprises of medium- to fine-grained sandstone exhibiting unidirectional cross-stratification, having average set thickness 25 cm. Internally, the cross-stratifications are characterized by alternation of thick–thin planar tabular cross-strata, the foreset bundles are separated by mudstone partings, thickness measurement of cross-set bundles reveals a near symmetrical pattern of cyclicity. A maximum of 28 laminae occur between two successive peaks of the thick laminae	The internal structures of sandstone, cyclicity in laminae thickness and concentration of mud within all the facies indicates tide-dominated depositional setting. The cyclicity measured from alternating thick–thin laminae is comparable with the lunar bimonthly (spring to spring) cycle
Herringbone cross-stratified sandstone facies	Two sets of oppositely dipping cross-strata separated by a gently inclined, planar erosional surface constituting a herringbone pattern (average thickness is 30 cm). The orientation of herringbone cross-strata shows distinct bipolar and bimodal paleocurrent direction. Double mud drape is a characteristic feature present within the foresets	Bipolar and bimodal paleocurrent direction of herringbone cross-stratified sandstone facies corroborates tidal actions. The presence of double mud drapes within this facies indicates a subtidal environment
Ripple-laminated facies	Characterized by small-scale ripple-laminated sandstone, mud is present at the trough of some of the ripples, and maximum foreset thickness is 5 cm. Thick mud partings occur at the ripple set boundaries, and cross-lamina exhibits sigmoidal pattern at places	Must have been deposited during the waning phase of a tidal current
Planar-laminated facies	Planar-laminated sandstone with intermittent mud laminae defines this facies. Average thickness of the facies is 20 cm. It exhibits vertical variations in lamina thickness (ca. 2 cm for sand lamina and 0.3 cm for mud lamina)	Likely to be deposited during slackening period of the tidal current

(continued)

Table 2.8 (continued)

Facies	Description	Interpretation
Conglomerate facies	This facies occurs locally and is characterized by matrix-supported conglomerate unit having wedge-shaped geometry with maximum thickness of ~40 cm. The base of the facies is invariably sharp and scoured. The compositions of clasts include chertified limestone, vein quartz and sandstone. Maximum size of the clasts is 11 cm. The interstitial spaces between the clasts of the conglomerate are filled up by coarse-grained sand matrix. This conglomerate facies often grades into massive sandstone both laterally and vertically	The vertical and lateral transitions of conglomerate to massive sandstone indicate deposition of both the components from a single flow. The scoured base of the conglomerate beds at places indicates the presence of turbulence within the flow
Tabular cross-stratified, amalgamated sandstone facies	This facies is characterized by amalgamated sandstone beds. Maximum thickness of individual bed is of ca. 32 cm but amalgamation may contribute a thickness up to 1 m. Internally tabular cross-stratifications define these amalgamated sand beds	Occurrence of amalgamated sandstone beds, juxtaposed one above other, suggests rapid recurrence of an event flow, possibly storm surges, in a high flow regime
Hummocky cross-stratified sandstone facies	This facies is characteristically medium- to fine-grained, moderately sorted sandstone with broadly lenticular to tabular beds with convex-up tops. It is overlain by wave ripple-laminated sandstone with less sharp contact. Hummocks and swales are frequently observed with maximum height and wavelength of ca.8 cm and 35 cm, respectively	Likely to be deposited by a storm current-induced flow
Wave ripple-laminated sandstone facies	This wave ripple-laminated sandstone facies generally overlies the top of the preceding facies. Ripple crests are straight and show bifurcations on the bedding plane. Syneresis cracks are abundantly present on bed surfaces at different levels. Overall orientations of these cracks are fairly consistent but vary widely between beds	Numerous wave features on the sandstone bed surface indicate the facies is product of wave-dominated flow

(continued)

Table 2.8 (continued)

Facies	Description	Interpretation
Shale facies (Bijaigarh Shale)	Composed of black colour, sulphidic shale. Thickness of this facies increases towards the eastern part of the Son valley. Maximum thickness occurs at Amjhore. Thick mineralization of pyrite presents within this facies in this area (ca. 3 m). In the lower part of the shale, thin sandstone band and gutter filling sands are present but in the upper part it is almost devoid of sand. Organic carbon content of this shale frequently exceeds 3%. The top part of this facies is marked by the occurrence of 10-m-thick white siliceous claystone	The facies is likely to be deposited in shelf area. The upper sand-free shale must have been deposited in the distal outer shelf environment where sand and silt could not be delivered even during storm. Anoxic condition is indicated by high organic carbon content and the occurrence of syngenetic pyrite

Fig. 2.21 Vertical outcrop section of the Scarp Sandstone (width of photo = 130 m) near Amjhore

fine sandstone characteristically constitute ca 42-m-thick Lower Bhander Sandstone. However, its basal part attains a thickness of up to 11 m, consisting of grey shale with intervening grey sandstone. Salt pseudomorph occurs abundantly within the Lower Bhander Sandstone, and deposition of this sandstone must have been taken place in nearshore paleogeography. The Sirbu Shale overlying the sandstone is either a lagoonal (Singh 1973) or a shelfal (Sarkar et al. 2002b) deposit. The youngest stratigraphic unit, the Upper Bhander Sandstone Member, overlying Sirbu Shale, represents an eolian–fluvial deposit with supralittoral sediment confined to its base (Bose et al. 1999).

Table 2.9 Description and interpretation of the facies constituting the Scarp Sandstone Member around Churk–Mirzapur area

Facies	Description	Interpretation
Storm- generated facies	This facies is characterized by metre-scale parallel-sided, amalgamated sandstone beds grading upwards into laminated shales. The beds are laterally continuous. Gutters and flutes are present at the base of the sand beds. Mud clasts often occur at the bottom of the sandstone beds. Chakraborty and Bose (1992) and Chakraborty (1993) identified sedimentary structures within these beds that are similar to that of Bouma sequence. Massive sandstone with mud clast concentration often present at the base of the sequence is followed by mm-thick parallel laminations. Ripples with straight crests are common at the top part of the sequence. Hummocky cross-stratifications, though rare, are reported from the sandstone	As evident from the description, the deposition had taken place from suspension under the influence of strong current. The common bed top ripples with straight cress are likely to be of wave origin. All these features along with the sole features suggest that the sand beds were deposited by storm current. Occasional presence of hummocky cross-stratification also supports the presence of storm currents

2.9.1 *Ganurgarh Shale*

The Ganurgarh Shale, the lowermost member of the Bhander Formation, appears reddish grey in outcrop and is composed of alternation between fine sandstone and mudstone layer with plenty of ripples on the bed surface (Figs. 2.26, 2.27). The overwhelming dominance of combined flow structures, wide paleocurrent dispersal, the red colour, and the local mud cracks and halite crusts in combination with the gradational transition into the shoreface carbonate of the Bhander limestone indicate marginal marine paleogeography for the Ganurgarh Shale (Chakraborty et al. 1998). Calcarenite sheets replace the quartz sandstone facies at the top part of the Ganurgarh Shale. The Ganurgarh Shale is constituted by three distinctive facies reflecting different sedimentation conditions. The description and interpretation of the facies are given below (Table 2.13).

2.9.2 *Bhander Limestone*

The entire Bhander Limestone is well exposed in quarry sections around Dolni, Girgita and Emilia around Maihar and Babupur, Sajjanpur and Sagmanhia around

Table 2.10 Description and interpretation of facies constituting the Dhandraul Sandstone Member around Churk–Mirzapur area

Facies	Description	Interpretation
Subaerial facies assemblages		
Flat, parallel-laminated sandstone	Laterally persistent, tabular in geometry, thinly (3–4 mm) parallel-laminated sandstone, parting lineation on bedding plane, occasional presence of channel-like scours at sandstone base	Considering the primary structures, textures, geometry of the sand body, channel lag on the planar erosional surface, this part of the Dhandraul Sandstone is likely to be a product of braided ephemeral rivers. The presence of cosets of parallel-laminated sandstone with parting lineation, cosets of trough cross-bed sets and erosion surfaces covered with intraformational conglomerate reflects the varying flow stages of the streams
Large-scale, low-angle cross-stratified sandstone	Characterized by thick cross-stratified sets, wedge-shaped, straight to slightly concave-up cross-strata with tangential toe, foreset inclination is low, with average set thickness 60 cm. Individual set is separated by sharp planar erosional bounding surfaces covered with intraformational conglomerate. Foresets are long and asymptotic, with low foreset slope (10–15°). Deformed foreset (overturned) is occasionally present. Downstream accreted macroform is locally present	
Trough cross-stratified sandstone	Small-scale trough cross-stratifications (width varies from 10 to 15 cm), cosets thickness varies from 20 to 150 cm, capped by laterally discontinuous thin (few cm) mudstone/siltstone layer	
Oppositely dipping cross-stratified sandstone	Characterized by cosets of cross-strata both planar to trough variety, thickness of cross-strata sets varying from 0.3 to 0.8 m. Tabular to lenticular in geometry, many exhumed dune bedforms are present with prism-shaped morphology. Height of the bedform varies from 0.3 to 1.5 m, and their width ranges from 2 to 5 m (Chakraborty 1993). Dunes are composed of inclined cross-strata. In transverse section, the strata show zigzag pattern	Longitudinal dunes possibly formed by deflection of repeated oscillation of wind across dune crest (Chakraborty 1993). Zigzag pattern of the stratal arrangement suggests vertical accretion of longitudinal dune, possibly formed within an eolian domain

(continued)

Table 2.10 (continued)

Facies	Description	Interpretation
Subaqueous facies assemblage		
Parallel-laminated sandstone	Represented by parallel-laminated sandstone with parting lineation, extensive in occurrence	The parallel-laminated sandstone with parting lineation might have been deposited in shallow marine high energy environment, may be a beach. On the other hand, occurrence of herringbone cross-strata at places evidently indicates reversing tidal currents in a tide-dominated shelf
Herringbone cross-bedded sandstone	Sandstones show herringbone cross-stratifications. The two sets of unidirectional cross-stratifications are separated by erosional surfaces. The thickness variation within the individual bundles reveals alternate thick and thin lamination	

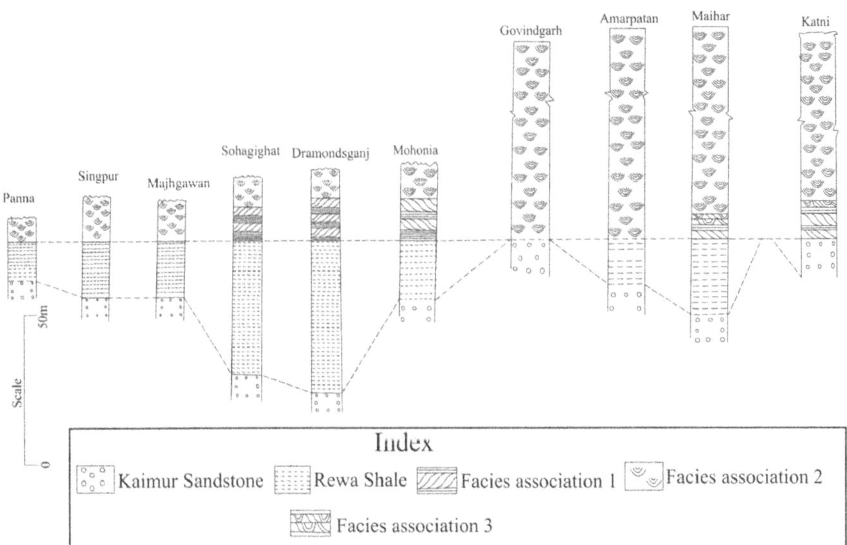

Fig. 2.22 Panel showing lateral and vertical facies variation of the facies association constituting the Rewa Formation

Satna. Facies sequence across the study area indicates several transitions between offshore and nearshore domains, indicating alternate rise and fall of relative sea level (Table 2.14, Fig. 2.28; for details, see Sarkar et al. 1996). Stromatolite morphology reflects distinctive changes, reflecting transitions from relatively deep to shallow marine facies (Fig. 2.29). The dominance of wave features, prominent westward slope and depth-related facies organizations indicate deposition of carbonate sediments on an open ramp setting.

Fig. 2.23 Characteristic features of Rewa Shale: granular lag (**a**), planar-laminated shale (**b**), gutter casts under a sandstone bed (**c**), an isolated gutter, filled by sandstone (pen length = 14.5 cm)

Table 2.11 Description and interpretation of facies constituting the Rewa Shale

Facies	Description	Interpretation
A	Consists of black to greenish-black shale. The sandstone/siltstone interbeds are not laterally persistent. Intertwined light and dark sublaminae (2–5 mm thick) within shale; sand/silt interbeds are massive. Total organic carbon content of the shale is usually less than 1%	Deposited within outer shelf beyond wave base
B	Consists of grey shale with sandstone sheets. The sandstone beds exhibit hummocky cross-stratification (HCS) and planar lamination. Organic carbon content of the shale is ~0.5%	Deposited in inner shelf, and within the storm wave base
C	Reddish-brown shale interbedded with rippled sandstone and HCS sheets. Sigmoidal, long-toed, low-angle asymmetric ripple laminae. Thicker sandstone units are either channelized or tabular, with planar and wavy laminae. HCS occurs locally. Organic carbon content is ~0.1%	Distal shoreface deposits

Table 2.12 Description and interpretation of the facies association comprising the Upper Rewa Sandstone

Facies association	Description	Interpretations
Marginal marine facies association		
Large-scale unidirectional cross-stratified facies	It is characterized by large-scale unidirectional planar cross-stratification (80 cm–2 m set thickness). The style of cross-stratification changes laterally from tabular to sigmoidal to concave-up within discrete successive package. The package boundary is defined by erosional surface, generally parallel with respect to the foreset. The foreset commonly bears parting lineation. Within a package, the lamina thickness shows an initial decreasing and then an increasing trend. The number of laminae constituting the package varies from 22 to 28. Locally bimodal, bipolar paleocurrent has also been observed. At places on the foreset of unidirectional cross-stratifications, small-scale ripples are present migrating upslope. Millimetre-thick mud veneers protect the ripple forms. Reactivation surfaces are frequently present	The unidirectional large-scale cross-stratification formed presumably by migration of sand waves. The ripples migrating upslope on the foreset of cross-stratification indicate current reversal and indicate tidal current. Frequent presence of reactivation surfaces supports this contention. The reverse orientations between the large-scale cross-strata and the rare mud-draped ripples on them record intermittent flow reversals, suggesting strong tidal asymmetry. The package having 22–28 laminae may be the product of semidiurnal tide
Plane-laminated facies	The plane-laminated units are subordinate in occurrence and present in close alternations with earlier facies. Their thicknesses range from 60 to 90 cm. The bedding plane surface is carved with parting lineation and current crescents. Orientations of these bed surface structures are the same	The plane laminations with parting lineation and current crescents presumably formed by strong shear under tractive or oscillatory flows. Alternations between the two facies, each having substantial thickness may possibly a reflection of alternate domination of tide and wave. The rhythmic pattern of cross-stratifications along with beach facies support tidal interpretation

(continued)

Table 2.12 (continued)

Facies association	Description	Interpretations
Fluvial facies association		
Cross-stratified granular sandstone	This sandstone facies, often granular, characteristically forms lenticular solitary bodies of thickness up to 90 cm. They are internally planar cross-stratified, resting on planar or broadly undulated master erosion surfaces. The foresets are commonly low dipping and granules when present, demarcate the foreset bases. Foresets are normally graded in nature. Slump structures and overturned cross-stratification are abundantly present	Sandstone of this association is generally texturally immature in terms of grain sorting, often incorporating granules, even small pebbles. Fine-grained and well-sorted sandstone occurs locally. The coarse-grained trough cross-stratified sandstone, with overturned cross-stratification possibly deposited in a braided fluvial regime. The well-sorted fine-grained sandstone in vertical and lateral continuity with this sandstone seems to be of eolian patches deposited within the braided fluvial channel deposit
Trough cross-stratified sandstone	Generally, granule-free, medium-grained trough cross-stratified, lenticular sandstone bodies. The cross-set thickness is ca 15 cm and average foreset dip is 14^0. The coset thickness varies from 18 cm to 1 m	
Down dip cross-stratified sandstone	This is characterized by compound cross-stratification, major cross-strata dipping at about 2–3°, and the smaller internal planar cross-laminae make an angle of about 17–19° with them in the same direction	
Cross-stratified granule	This facies contains the coarsest grain within the Upper Rewa Sandstone. Its granule content is more than 30%. It is very poorly sorted, yet thoroughly cross-stratified	
Thinly laminated medium- to fine-grained well-sorted sandstone	Well-sorted, medium- to fine-grained and thinly laminated sandstone showing tabular geometry, erosionally terminated laterally by some other facies. The internal laminae characteristically tend to be horizontal maximum dip of about 10^0. Internally, they are preserved as ripple forms or cross-laminations at places. At times low-angle planar strata of thickness about 10 cm are present; significantly, these strata show distinct concentration of coarsest material on top of them	

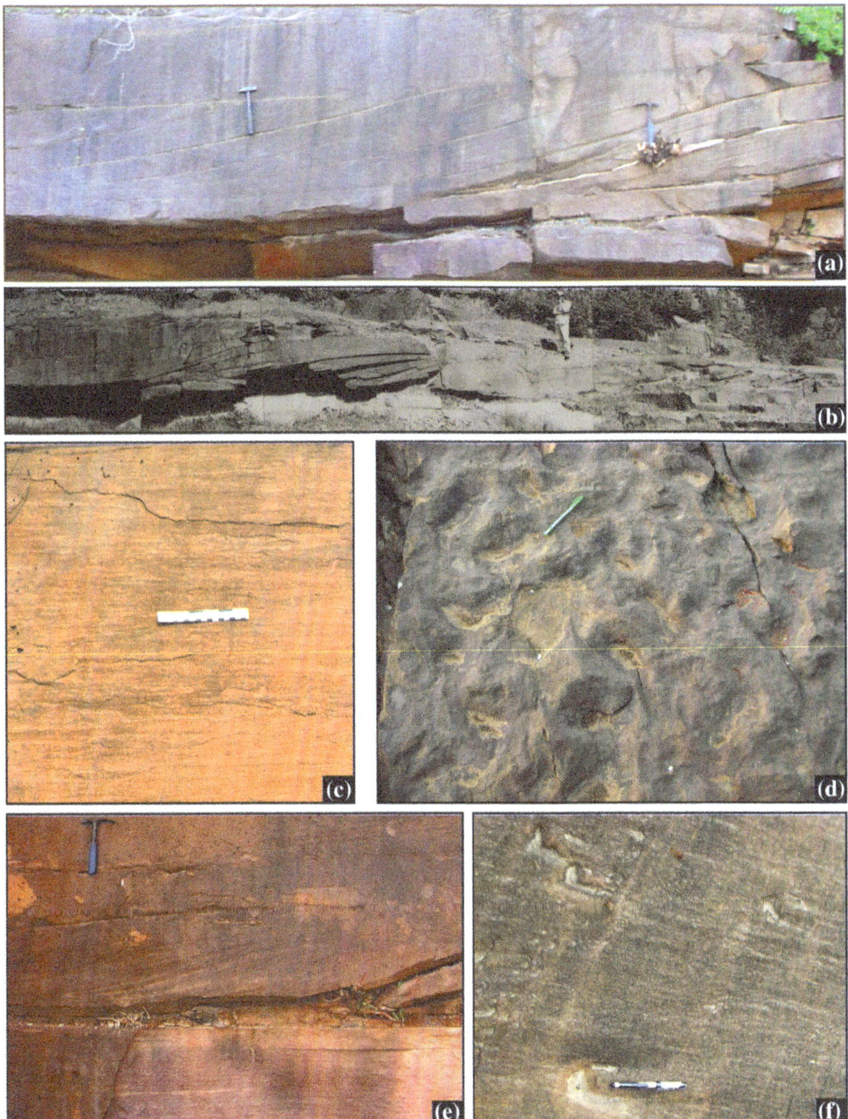

Fig. 2.24 Photographs showing features in the marginal marine facies of the Rewa Sandstone: large-scale unidirectional planar cross-stratification (**a**), tabular to sigmoidal transition of cross-stratification (**b**), parting lineation on a sandstone bed surface (**c**), small-scale ripples (**d**), planar and cross-laminated sandstones in profile (**e**) and current crescents on a sandstone bed surface (**f**) (pen length in d, f = 14.5 cm, hammer length in a, e = 38 cm)

Fig. 2.25 Fluvial facies association within the Upper Rewa Sandstone: cross-stratified granular sandstone (**a**) slump folds underlying cross-stratified sandstone (**b**) overturned cross-stratification (**c**), trough cross-stratified sandstone (**d**), compound cross-stratification (**e**), cross-stratified granule bed (**f**), thinly laminated sandstone (**g**) and ripple laminae (**h**)(hammer length in a, b, c, e, g = 38 cm; pen length in d = 14.5 cm; matchstick length in h = 4.5 cm)

2.9.3   *Lower Bhander Sandstone*

The Lower Bhander Sandstone overlies the Bhander Limestone. The outcrops of the Lower Bhander Sandstone are limited because of the high mud content. Well-exposed sections of the Lower Bhander Sandstone occur near Dolni, and Lilji and Patna nullahs in Maihar (Figs. 2.30, 2.31). The ~42-m-thick Lower Bhander Sandstone primarily consists of red mudstone and fine-grained sandstone (Table 2.15; Fig. 2.30).

The lower part of the Lower Bhander Sandstone represents a prograding shelf. The upper part indicates deposition in a coastal playa. In this microtidal setting, mud deposition takes place during the fair weather time, while sand deposition takes place

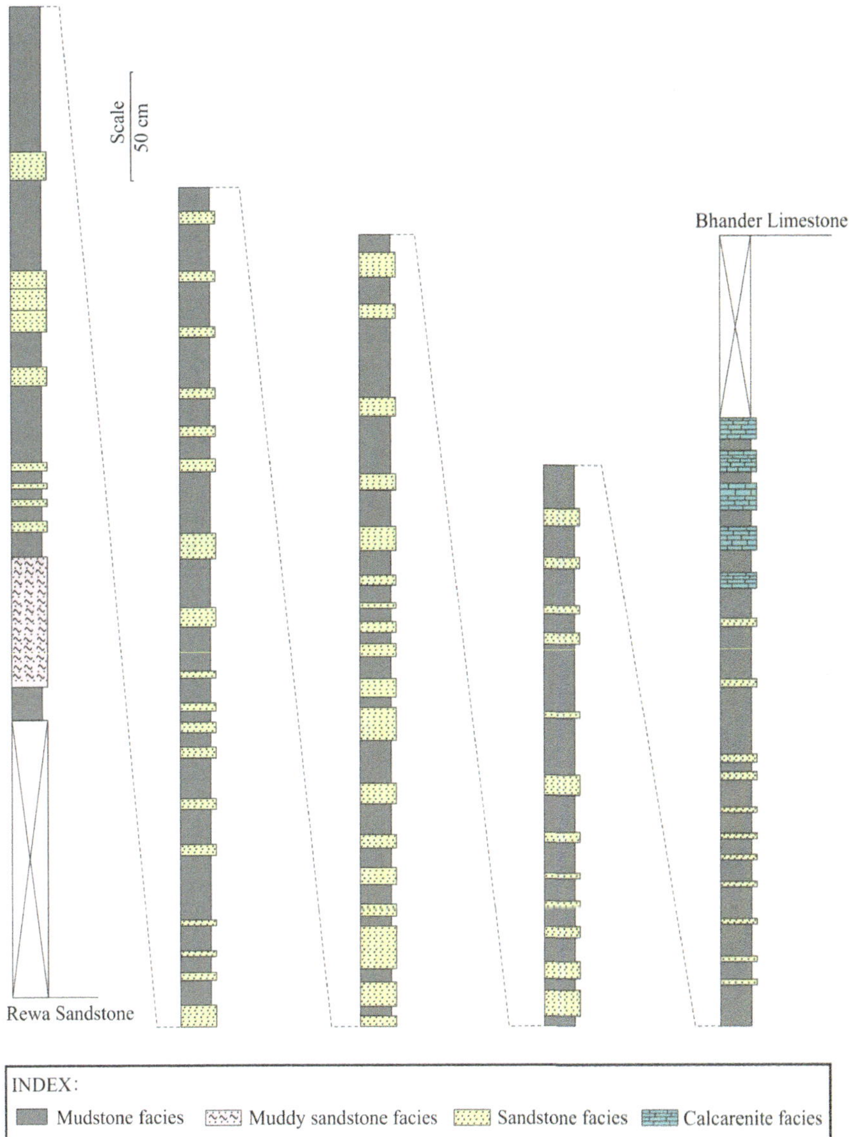

Fig. 2.26 Litholog showing constituent facies of the Ganurgarh Shale

Fig. 2.27 Planar-laminated (**a**) and ripple-laminated mudstone (**b**) within the Ganurgarh Shale (pen length = 14 cm, hammer length = 38 cm)

during storms. Resting on the shelf-originated carbonates of the Bhander Limestone, the Lower Bhander Sandstone records a small-scale regression.

2.9.4   Sirbu Shale

The Sirbu Shale, overlying the Lower Bhander Sandstone, contains stromatolitic and oolitic patches alternating with grey shale at the lower part. This part has been interpreted as a lagoonal deposit (Singh 1973) followed by entirely siliciclastic shelf deposits. An organic-rich, sand-free shale, devoid of emergence features overlies the lagoonal shale, intervened by a cm-thick conglomerate and sandstone beds. The shelfal Sirbu Shale exhibits an overall coarsening-upward trend, incorporates more storm originated siltstone and sandstone beds upwards with abundant bipolar sole marks and gradationally passes upwards into the Upper Bhander Sandstone (Sarkar et al. 2002b).

The Sirbu Shale comprises two facies assemblages: mixed siliciclastic–carbonate assemblage and siliciclastic assemblage. A mixed siliciclastic–carbonate facies assemblage marks the basal part the Sirbu Shale that gradationally overlies the Lower Bhander Sandstone (Figs. 2.32, 2.33). The basal part of Sirbu Shale is best exposed around Lilji and Patna Nala near Maihar, and it attains a thickness of ca. 7.2 m. A summary of the constituting facies of the mixed siliciclastic–carbonate assemblage is

Table 2.13 Description and interpretation of facies constituting the Ganurgarh Shale around Satna–Maihar area

Facies	Description	Interpretation
Mudstone facies	The mudstone facies is massive or exhibit parallel to subparallel, mm-scale fine sand/silt inter-laminae and intercepted by very low-angle planar erosion surface. Isolated layers with asymmetric sand ripples of amplitude 0.6 cm and wavelength 2 cm, lacking internal lamination occur locally. Locally, the cross-laminae show chevron arrangement. On the bed surface, ripple crest are straight or slightly sinuous with bifurcation	The mudstone facies with very small ripples of combined flow origin clearly reflects attenuation of web and or tide energy, possibly due to the dampening of a muddy substrate. The alternations of mud and silt with sand laminae can be explained by frequent fluctuation in depositional energy
Sandstone facies	It alternates with mudstone facies and consists of fine sandstone. Three-dimensional reconstructions reveal their ridge-like geometries. The sandstone beds have their bases sharper than their tops depicting overall grading. Internally, they are characterized by planar or low-angle lamination intercepted by broadly, wavy erosion surface. Chevron cross-stratifications are locally present. Combined flow wave-current ripples with straight and locally bifurcated crests are abundantly preserved on the bed surface	The sandstone facies dominated by plane lamination indicate to a temporal steep rise in depositional energy. The ripples of combined flow origin indicate both the presence of current and wave. The current may be tidal
Muddy sandstone facies	This facies is present only at the base of Ganurgarh Shale. It consists primarily of fine sandstone, but closely intercalated with mud flaser and thin laterally persistent mud laminae	Indicate frequent fluctuation in flow conditions in a tidal environment

Table 2.14 Description and interpretation of facies constituting the Bhander Limestone around Satna–Maihar area

Facies	Description	Interpretations
Shale facies	It consists of thin interbeds of siltstone and very fine sandstone alternating with dominant shales. The shales are plane-laminated and exhibit greenish-grey colour. The lower boundaries of the coarser clastics are invariably sharper than their tops. Internally, these beds exhibit wave ripple lamination	The plane-laminated greenish-grey shales indicate deposition in quiet and sub-oxic condition. Sharp base and gradational tops of siltstone/sandstone beds indicate episodic deposition. Wave ripple lamination indicates action of waves. Deposition of the facies possibly took place near the storm wave base
Plane-laminated mud facies	The facies consists of lime mudstone, and it exhibits mm-thick plane laminae. Each carbonate lamina is draped by a thin argillaceous veneer. Thick–thin lamina alternation is common in this facies. The carbonate laminae may show normal grading at places. Locally, it exhibits siltstone shale rhythmite	Fine grain size and planar laminae indicate low-energy depositional conditions. Calcareous laminae exhibiting thick–thin alternation indicates tidal deposition (Sarkar et al. 1996)
Heterolithic facies	It consists of fine calcarenite sheets alternating with either plane-laminated mudstone or grey shale. Calcarenites display various kinds of wave ripple laminations, unidirectional small-scale cross-stratifications and hummocky cross-stratifications	Alternations between calcarenite and mudstone indicate fluctuation in energy. Calcarenites beds are identified as storm deposits, while the mudstone beds are indigenous
Stromatolitic facies	Consists of two broad types of stromatolites, viz. stratiform and biohermal (Sarkar et al. 1996). The biohermal type is more abundant. The stratiform type may show occasional clusters of micro-columns and is similar to tufa stromatolites. Two broad types of biohermal stromatolites occur. The columns in first variety exhibit frequent branching and preferred orientation, while the columns are less branching in the second variety	Biohermal forms possibly indicate greater depth than biostromal variety (Beukes and Lowe 1989). Tufa stromatolites possibly indicate marginal marine setting (Grotzinger 1989). Branching with strong imbrications indicates strong current/wave actions. Lateral association with the planar and cross-stratified facies indicate shoreface depositional conditions for both

(continued)

Table 2.14 (continued)

Facies	Description	Interpretations
Planar and cross-stratified facies	Consists of cross-stratified calcarenites. Cross-stratifications may be large or small (average set thickness 18 and 7 cm, respectively). The large sets are mostly planar, chevron-like or rarely herringbone type. Small sets are chevron-like, and they locally grade to planar lamination. Symmetric wave ripples occur locally. At places, the facies exhibits broad wavy laminations with hummocky and swaley cross-stratifications	The large scale of cross-stratifications and intercalated facies indicate shallow-water origin. The cross-stratification patterns indicate dominance of oscillatory flows. Uncommon herringbone cross-stratification indicates tidal influence. Hummocks and swales bear signature of storms

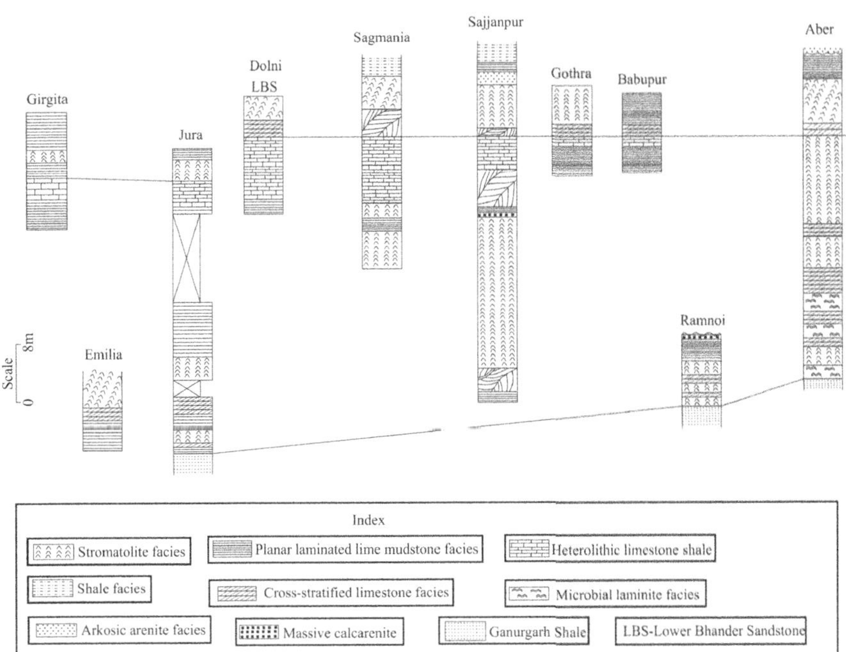

Fig. 2.28 Temporal and spatial facies variations present within the Bhander Limestone in different localities

Fig. 2.29 Salient features of the Bhander Limestone: interbeds of siltstone and very fine sandstone alternating with shale (**a**), coarse clastic sandstone beds (**b**), heterolithic facies (**c**), small-scale ripple (**d**), hummocky cross-stratification (**e**), inclined stromatolite (**f**), bioherm stromatolite (**g**), herringbone cross-stratification (**h**) (hammer length in a, c, d, g = 38 cm; pen length in b, f = 14.5 cm; Swiss knife length in e = 9.1 cm)

provided in Table 2.16. Dark-coloured shale, containing subordinate siltstone beds, sharply overlies the basal facies assemblage. The contact between the two facies assemblages of the Sirbu Shale is marked by a mud-pebble conglomerate.

The basal lagoonal part of the Sirbu shale passes upwards to an argillaceous shelfal part, devoid of emergence features (Fig. 2.34). The upper part of the Sirbu Shale is well exposed around Maihar area (Fig. 1.1). This part attains a maximum thickness of 185 m and it consists of six facies (Table 2.17). A detailed facies investigation reveals an overall prograding upward pattern in the Sirbu Shale (for details, see

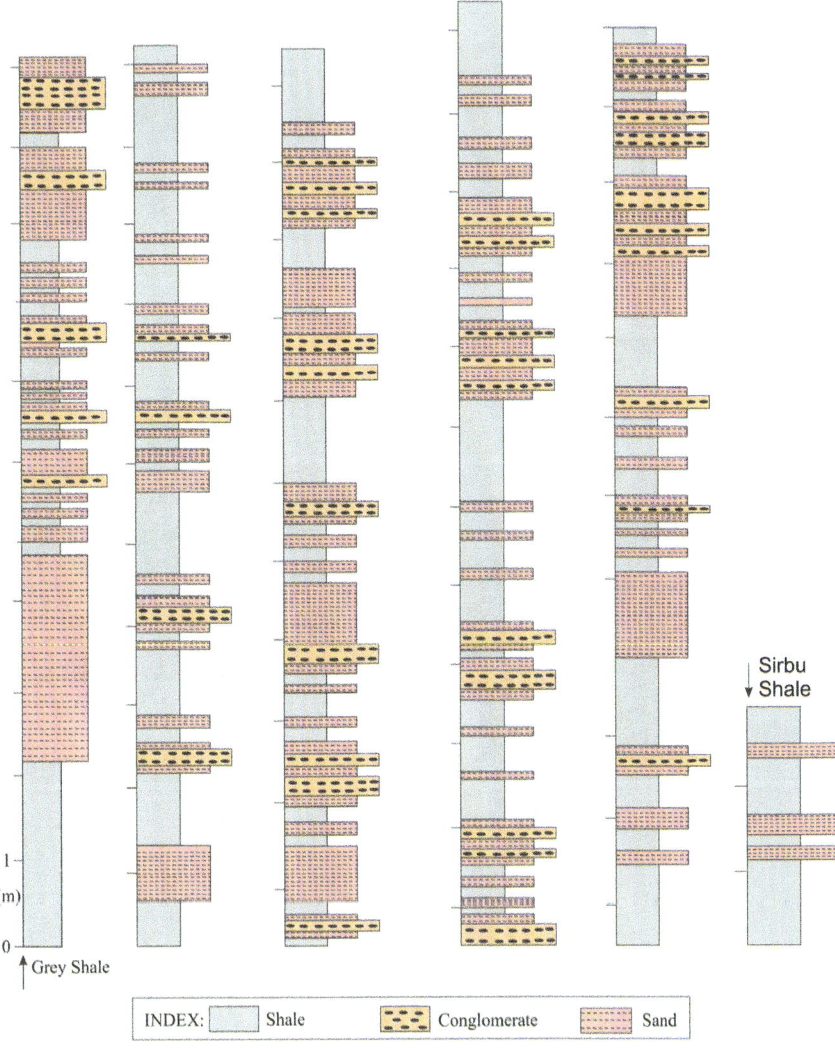

Fig. 2.30 Litholog showing variation of facies constituting the Lower Bhander Sandstone

Sarkar et al. 2002b). The upper section of the late Proterozoic Sirbu Shale indicates deposition on a storm-dominated shelf. Five of the six facies are laterally extensive, cyclical, consisting of interbedded shale–siltstone/fine sandstone. They have gradational mutual transitions (Fig. 2.35). The sixth facies consists of coarse sandstone of low textural and mineralogical maturity, and occurs locally, encased by deeper shelf facies (Fig. 2.33). Many of the coarse interbeds bear signatures of storm actions. Current-formed features below storm beds record dominantly shore-parallel flow, although a shore-normal component is also evident.

Fig. 2.31 Sedimentological features of the Lower Bhander Sandstone: conglomerates consisting of mud clasts (**a**), massive sandstone (**b**), tabular sandstone (**c**), sole marks under a sandstone bed (**d**), salt pseudomporhs on a sandstone bed (**e**), red mudstone alternating with sandstone (**f**), grey shale in facies A (hammer length in b = 38 cm; pen length in a = 14.5 cm; coin diameter in e = 2.3 cm, outcrop width in f = 1.5 m)

Fig. 2.31 (continued)

The characteristics of facies comprising the Sirbu shale are similar to storm-dominated shelf successions (e.g. Dott and Bourgeois 1982; Bose et al. 1988; Sarkar et al. 2002a, b). The consistent trend of the wave ripple crests suggests an N–S paleoshoreline alignment. Paleocurrent data suggest storm-induced flows were dominantly shore-parallel geostrophic, although shore-normal particle movement also occurs. Facies A appears to be distal-most offshore facies, but bipolar sole marks indicate its deposition above the storm wave base. The wispy, darker laminae within facies B and C possibly represents microbial mats. Bipolar prod marks, wave ripples and hummocky cross-stratification indicate an oscillatory component of the flow. Transitions from facies A to E through B, C and D are marked by the increasing lateral continuity, grain size, thickness and bedform dimensions of coarser components. The transition from facies A to E indicates increasing proximity to the shore. The presence of wave ripples within the relatively finer interbeds in facies D and E suggests deposition within fair weather wave base, whereas facies A, B and C were deposited largely between fair weather and storm wave base. The facies F with the undulatory base, internal hummocky cross-stratification and wave ripples indicates the action of storm waves of distinctly greater intensity. Earthquakes could have induced the waves. The sharp erosional base creating alternate domal hummocks and bowl-like swales indicate pre-depositional wave erosion. The basal low-angle

Table 2.15 Description and interpretation of the constituent facies of the Lower Bhander Sandstone around Maihar area

Facies	Description	Interpretations
Conglomerate grading into sandstone	The facies consists of conglomerate and massive sandstone, both exhibiting wedge-shaped geometry. The conglomerate often grades to sandstone both laterally and vertically. Most of the mud clasts within the conglomerate are flat. The conglomerates are matrix-supported. The clasts are haphazardly oriented, occasionally showing incipient grading. Bases of the conglomerates are always sharp, but their tops are less sharp. Wedge-shaped bodies exhibit striations on bevelled edges. These slide planes are often exhumed and vertically juxtaposed	Wedge-shaped geometry and haphazard orientation of the clasts within matrix-supported conglomerates suggests their debris flow origin. Vertical and lateral transitions to sandstones indicate deposition of both components from a single flow. Association of slide planes suggests that the debris flows are possibly triggered by slides
Tabular sandstone	It consists of tabular fine-grained sandstone beds with localized mud inter-laminae. The bases of the sandstone beds are sharp, but tops are gradational with the overlying red mudstone. Beds show upward transitional in structural elements: a massive- or planar-laminated part with a basal lag of mud clasts gives way upwards to meso-scale cross-stratification and finally to ripple drift laminae. Halite pseudomorph occurs profusely at the bases of sandstone beds generally, but some are present on the bed surface as well. Flutes and tool marks occur at the soles of sandstone beds. Bed geometry of the cross-stratifications changes cyclically showing tidal bundle	The abundance of wave ripples and emergence features indicate coastal shallow-water deposition. Abundant sole features indicate supercritical nature of the sand-laden flow. The steady upward decline in flow regime and grain size indicates progressive waning nature of the flow. Cyclicity within the cross-stratified division indicates alternate decrease and increase in a unidirectional flow, may be a storm-tide interactive system

(continued)

Table 2.15 (continued)

Facies	Description	Interpretations
Red mudstone	It consists of monotonous, massive to finely laminated reddish-brown mudstone alternating with tabular fine sandstone beds. Both mudstone and sandstone exhibit polygonal desiccation cracks. Kinneyia structures locally occur on top of sandstone bed surfaces	Abundance of desiccation features indicates possible intertidal depositional setting with occasional emergence
Coarse-grained sandstone	It consists of coarse-grained and poorly sorted, arkosic sandstone interbedded with thin grey shale. The facies occurs locally. The lower part of sandstone beds appears massive, planar-laminated at the upper part and wave-rippled at top. The sandstone beds thicken upwards at the cost of shale. The interbedded grey shale is similar to that of facies A. The shale is absent and sandstone is thoroughly cross-stratified at the top part	Coarse grain size and mineralogical immaturity of sandstones indicate close proximity to the shore, may be deposited by coastal storm
Grey shale	It consists of grey shale with lensoid yellowish fine sandstone/siltstone, mostly filling isolated gutters. The shale is parallel-laminated. Prod and groove casts occur at the soles of gutters. These are aligned parallel to the axis of gutters. The bases of the sandstone beds are always sharp and are marked by concentration of mud clasts. Gutter filling sandstones are generally planar-laminated near the base followed upwards by hummocky cross-stratification (HCS). Wave ripples on top of the sandstone beds are straight/sinuous crests and exhibit local bifurcations	Depositional energy remains low allowing suspension settling of mud. The presence of HCS, wavy, broadly concave-upward erosional surfaces and wave ripples indicate oscillatory component in the sand-laden flow. Sand emplacement takes place by high energy flows like storms. The facies appear to be a product of storm-infested inner shelf. The upward coarsening nature of the facies suggests increasing proximity to the shore

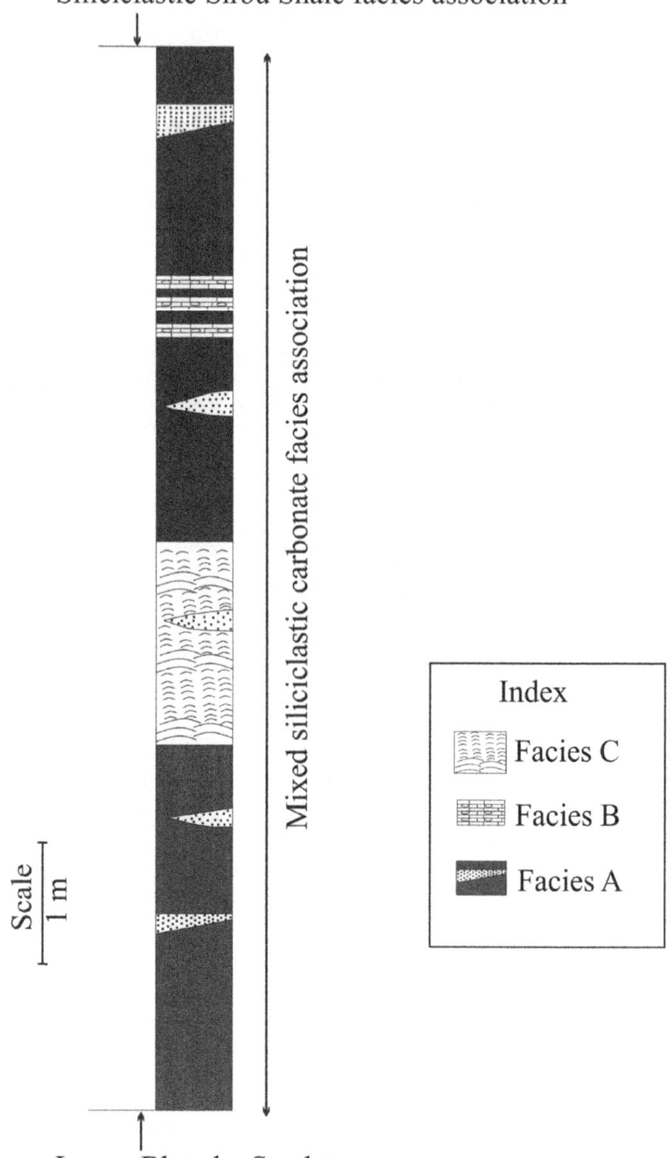

Fig. 2.32 Litholog showing constituent facies in mix carbonate–siliciclastic deposit of Sirbu Shale

Fig. 2.33 Stromatolite within Sirbu Shale (hammer length = 38 cm)

draping sand laminae are essentially hummocky and swaley cross-stratifications of Walker and Plint (1992).

2.9.5 *Upper Bhander Sandstone*

The Upper Bhander Sandstone is the topmost member of the Proterozoic Vindhyan Supergroup. The sandstone is well sorted and predominantly reddish brown in colour. It conformably overlies the marine, grey-coloured Sirbu Shale and is continuously exposed along the ridges, from Rampur to Hardua (Figs. 2.36, 2.37). The thickness of Upper Bhander Sandstone in the Son valley remains approximately the same throughout the stretch, about 100 m. The sandstone is petrographically quartz arenite, containing less than 5% feldspar and minor amounts of mica and rock fragments such as siltstone, chert with very negligible amounts of mica schist and phyllite. Despite the monotonous lithology, the UBS can be subdivided into three distinct

Table 2.16 Description and interpretation of the constituent facies of the lower part of the Sirbu Shale around Satna–Maihar area

Facies	Description	Interpretations
Stromatolite	Consists of two broad types of stromatolites, stratiform and biohermal. The former variety exhibits laterally linked hemispheroids, similar to tufa stromatolites. The hemispheroids are discontinuous and are interbedded with grey shale. The second variety occurs as flat-topped channel-filled bodies (av. width 2.5 m, av. thickness 1.2 m). Individual columns are club-shaped. Columns are arranged radially. Height and head diameter of columns vary 9–51 cm and 18–28 cm, respectively	Larger biohermal stromatolites indicate relatively deeper part of the marine environment, possibly in shallow channels within the lagoon with limited wave action
Oolitic limestone	The grey shale is intercalated with oolitic and intraclastic limestone at the mid-level. Broadly two varieties of oolitic limestones are found, viz. oomicrite and ooid-bearing intrasparite. The intraclast content may be up to 50% in the second variety	Association with grey shale indicates marginal marine environment. Possibly indicate deposition in a lagoon
Grey shale	It consists of grey shale alternating with shallow channel fill siltstone and fine-grained sandstone. It occurs immediately above the Lower Bhander Sandstone. The shale is fissile and it exhibits desiccation cracks. The siltstone/sandstone interbeds exhibit planar laminae and may contain solitary cross-sets in places. The facies may locally incorporate carbonate bodies	Abundance of polygonal desiccation cracks and its occurrence immediately above the salt pseudomorph-bearing Lower Bhander Sandstone indicates lagoonal origin

facies on the basis of limited variation in lithology, body geometry, structure, intrabed stratal organization and association. The facies are repetitive in occurrence in the stratigraphic column. The constituent facies of the Upper Bhander Sandstone are briefly presented in Table 2.18.

2.10   Sequence Stratigraphic Framework of the Vindhyan Supergroup

Bose et al. (2001, 2015) and Banerjee et al. (2006a) presented a detailed sequence Stratigraphic framework of the Vindhyan Supergroup. An unconformity and its laterally correlative conformity divide the Vindhyan Supergroup into two sequences. Each of the sequences is punctuated by three maximum flooding surfaces (MFS)

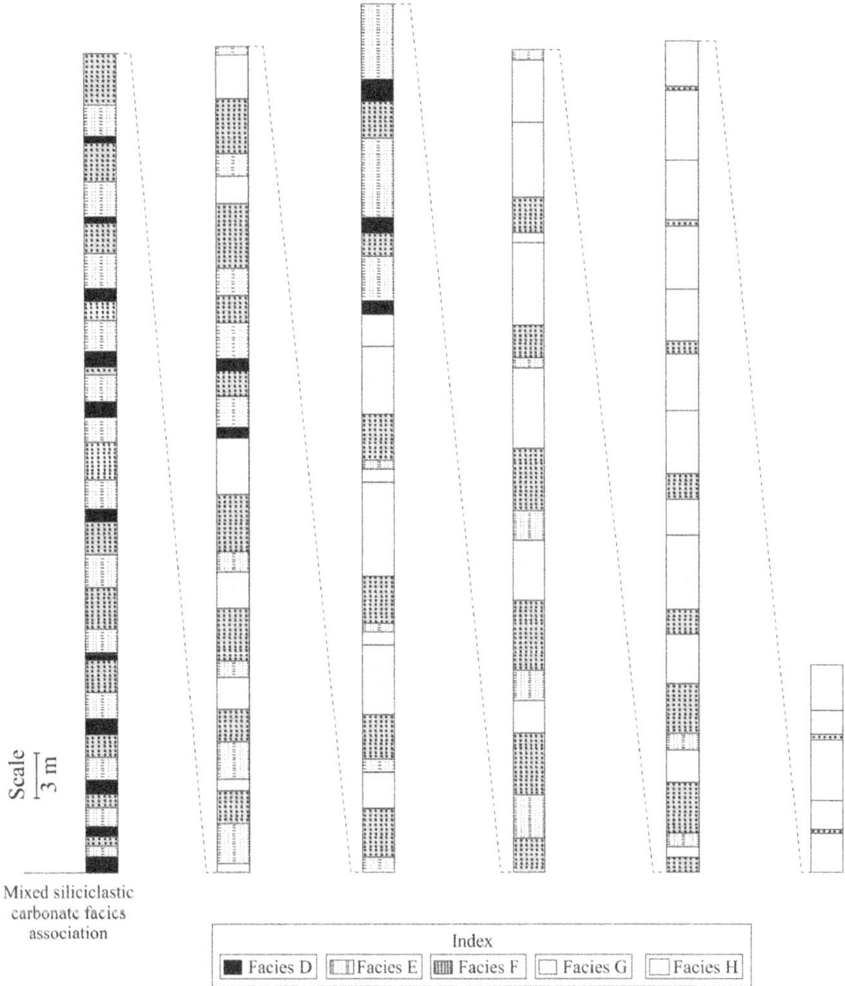

Fig. 2.34 Stratigraphic succession showing constituent facies variation within the siliciclastic part of the Sirbu Shale

(Fig. 2.38). Black shale, often pyritiferous, is associated with all MFS, which is considerably thick within the Rohtas Formation.

2.10.1 Lower Vindhyan

The vertical transition of depositional facies succession reveals systems tracts comprising the Lower Vindhyan sequence (Banerjee 1997; Bose et al. 2001, 2015). This lower sequence is ~1200 m thick and it consists of five formations, Deoland,

Table 2.17 Description and interpretation of the constituent facies of the upper part of the Sirbu Shale around Maihar area

Facies	Description	Interpretations
Facies F (sandstone)	It consists of poorly sorted, coarse-grained sandstones. It forms an anomalous element within facies C (thickness ca 1 m) and is laterally traceable up to half a km near Beta village. It is encased within facies C, and both of its lower and upper contacts are sharp. Bed thickness is laterally variable ranging up to 12 cm. Every bed shows pronounced grading. The base is undulated due to the presence of alternate scours and mounds, resembling hummocks and swales. The swales are up to 25 cm deep and up to 1.6 m wide. Upwards the parallel-laminated set gives way to sets of wavy laminations, including hummocky cross-stratifications. Bed tops show multiple sets of sinuous-crested, asymmetric mega-ripples of average wavelength 14 cm. Some bed surfaces show large-scale wave ripples (width ca.19 cm, height ca. 4.5 cm)	Deposition by storm waves of distinctly greater intensity or tsunami
Facies E (sandstone–sandy siltstone interbedding)	It consists of laterally extensive sandstone beds that alternate with coarse siltstone. Base of sandstone contains relatively larger gutters up to 19 cm wide and 14.5 cm deep. Above the gutters, sediments may be massive or planar-laminated and ripple-laminated towards the top. Ripples on bed tops are asymmetric. Sandstone beds are laterally persistent, ranging in thickness up to 22 cm and often amalgamate. The silty interbeds incorporate thin wave-rippled lenticles of very fine sandstone. At the sole of sandstone beds, tool marks are parallel to the gutters	Storm deposits above the fair weather wave base
Facies D (sandstone–shale interbedding)	Maximum facies thickness up to 72 cm and consists of interbedded very fine-grained sandstone and silty shale. The silty shale beds have an average thickness of 14 cm and comprise minute (mm amplitude) ripples encased in thin stringers of mud. The sandstone beds are sheet-like and range in thickness from 10.5 to 15 cm. Gutters on the base of beds have widths and depths of 14 and 12 cm, respectively. Wave ripples (width 8 cm, height 4.5 cm) occur on top of the beds. Internally, the beds show wavy laminae and moderate- to high-angle cross-stratification. The lower parts of the beds may be planar-laminated or massive. Tool marks occur under the sandstone beds	Storm deposits close to the fair weather wave base

(continued)

Table 2.17 (continued)

Facies	Description	Interpretations
Facies C (shale–siltstone alternation)	Alternate shale and siltstone beds of comparable thicknesses comprise this facies, with a thickness of up to 56 cm. The shales have same internal characters as their counterparts in facies A and B. The siltstones, form persistent beds of about 9 cm in thickness and have very sharp bases. They are either massive- or planar-laminated in their lower parts and ripple-laminated above. Wave ripples on top of them are larger than those in facies B. Gutters present on the soles of beds and are generally asymmetric. The majority of prod marks are parallel to the gutters and have the same N–S trend	Storm-induced sedimentation below the fair weather wave base
Facies B (shale with minor interbedding of siltstone)	This facies attains a thickness of up to 45 cm and differs from facies A by more frequent siltstone beds. The siltstone beds may be massive or planar-laminated and wave-rippled at the top. Grading is conspicuous within them. Gutters and groove casts occur at the soles of the beds. Gutters (av. width 4 cm, av. depth 3 cm) and ripples are larger than their counterparts in facies A	Offshore deposition frequently intervened by storm
Facies A (shale)	Dark green shale (max. thickness ca. 22 cm) with thin silt inter-laminae. The shale laminae often incorporate mm-thick, darker and wrinkled shaley wisps. The facies also contains lenticular siltstone beds. Siltstone beds are mostly massive or planar-laminated and locally finely cross-laminated. Their bases are sharp and planar, while their tops may be gradational. Soles of the beds bear tool marks, including groove casts and prod marks, oriented in an E–W direction	Overall quiet water depositional setting interrupted by distal storm deposits

Kajrahat, Porcellanite, Kheinjua and Rohtas, from bottom to top in order of superposition (Fig. 2.38). A ~2.8-m-thick conglomerate bed at the base of the Deoland Formation has been considered as a glacier deposit (Dubey and Chaudhary 1952; Chaudhury 1953; Ahmad 1955,1958; Mathur 1954, 1960, 1981). The overlying part of the Deoland Formation, which is shelf originated, gradationally passes over to the Arangi Shale, which forms the lower part of the Kajrahat Formation. The Arangi Shale gradationally passes upwards into the stromatolitic Kajrahat Limestone, which is frequently dolomitic and exhibits emergence features at the top part (Fig. 2). The overlying Porcellanite Formation is up to ~500 m thick, consisting of volcaniclastic tuff, pyroclastic flow and surge deposits (Banerjee 1997; Roy and Banerjee 2002). While Srivastava (1977) recognized subaerial deposition, Banerjee (1997) recorded frequent subaqueous to subaerial transitions within the Porcellanite Formation. The overlying Kheinjua Formation begins with a ~12-m-thick dark shale of offshore origin. The shelf originated Koldaha Shale consists of sandstone and shale alternations; both frequency and thickness of sandstone beds increase upwards, with local

Fig. 2.35 Photographs showing characteristic features of the Sirbu Shale: poorly sorted massive sandstone (**a**), hummock and swell in profile (below the hammer) (**b**), wave ripple on bed surface of a sandstone (**c**), alternation between sandstone and siltstone (**d**), gutter cast marked by yellow arrow (**e**), alternation between shale and siltstone (**f**) and thick exposure of shale (**g**) (hammer length in b, d, f = 38 cm; pen length in a = 14.5 cm)

intraclastic breccia wedge (Bose et al. 1997, 2001; Banerjee 2000; Banerjee and Jeevankumar 2005; Samanta et al. 2016). The Koldaha Shale gradationally passes upwards into the Chorhat Sandstone. The latter comprises mainly amalgamated, shallow marine storm beds (Seilacher et al. 1998; Sarkar et al. 2006). The Chorhat Sandstone is sharply overlain by the offshore-originated Rampur Shale, comprising the lower part of the Rohtas Formation (Sarkar et al. 2002a). The Rampur Shale becomes richer in organic content as it fines upwards and incorporates tuffaceous deposits towards the top (Sarkar et al. 2002a; Banerjee et al. 2006a). The organic-rich shale alternates with thin limestone beds and gradationally pass over to the overlying Rohtas Limestone. The latter is progradational in nature and exhibits wavy laminations, hummocky cross-stratifications and nodular beds in places and wave ripples that attest to shelf deposition (Banerjee and Jeevankumar 2007).

The gradational transition from the Deoland Sandstone to the Arangi Shale indicates retrogradation of the shoreline and supports a transgressive systems tract (TST; e.g. Van Wagoner et al. 1990; Sonnenfeld 1996; Catuneanu 2006). The overlying

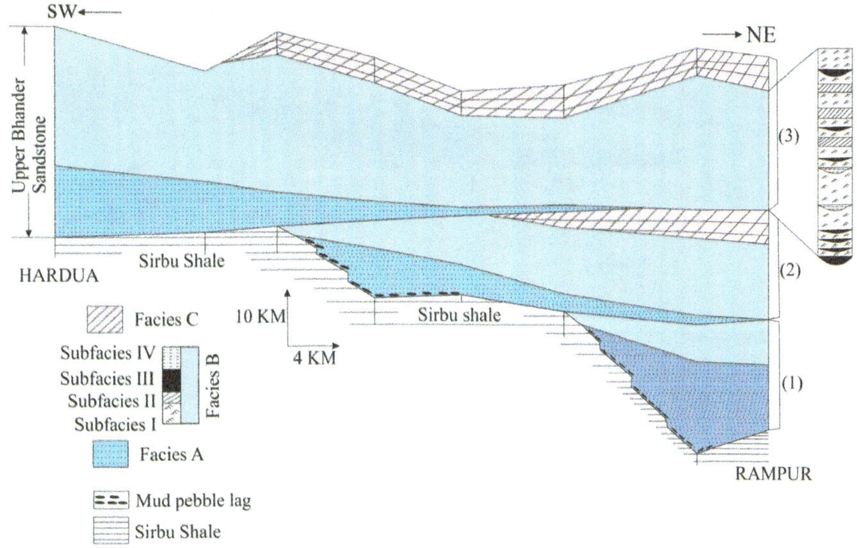

Fig. 2.36 Profile map from Rampur to Hardua showing distribution of facies/subfacies of the Upper Bhander Sandstone. Parasequences are numbered. Note stacking of three similar types of facies (1, 2, 3) with their bases partially demarcated by mud-pebble lags

Kajrahat Limestone is progradational in nature and thus indicates a highstand systems tract (HST; e.g. Sarg 1988; Van Wagoner et al. 1990; Catuneanu 2006). The maximum flooding surface (MFS; Galloway 1990; Sangree et al. 1990) most likely occurs at the base of the Kajrahat Limestone, containing organic-rich Arangi shale (Banerjee et al. 2006a). The overlying Porcellanite Formation consists of short shallowing-upward cycles and exhibits a slow rise in relative sea level (Banerjee et al. 2007). These cycles possibly result from the fluctuations of either, relative sea level or volcaniclastic deposition. The Porcellanite Formation has been considered as a TST (cf. Bose et al. 2001). The sand-free, organic-rich shale at the base of the Koldaha Shale indicates a rapid transgression and represents a condensed zone, the MFS present within it (e.g. Kidwell 1991; Gomez and Fernandez-Lopez 1994). The coarsening-upward Koldaha Shale and the Chorhat Sandstone together form an HST. The organic-rich Rampur Shale indicates another transgression. The Rampur Shale represents a TST, and a condensed zone occurs at its top. The Rohtas Limestone is overall progradational and is interpreted as an HST (Bose et al. 2001; Banerjee and Jeevankumar 2007). The Lower Vindhyan sequence, therefore, contains three MFSs (Table 1).

Fig. 2.37 Photographs showing characteristic features of the Upper Bhander Sandstone: alternating grainflow and grainfall laminae within eolian sandstone (**a**), thinly stratified sheet–sandstone (**b**), wrinkle structures on bed surface (**c**), isolated medium-scale cross-stratified sandstone (**d**), storm-laid mud clasts (**e**), large-scale cross-stratified sandstone (f), convolute-laminated sandstone bed (hammer length in d, f, g = 38 cm; Swiss knife in a, b, e = 9.1 cm)

Fig. 2.37 (continued)

2.10.2 *Upper Vindhyan*

The Upper Vindhyan sequence comprises three formations: Kaimur, Rewa and Bhander in ascending order of succession. Each formation begins with a shaley or calcareous segment and an upper sandstone segment. The shaley Lower Kaimur (locally known as Bijaigarh Shale) is lenticular in nature and is restricted to the southeastern sector of the Vindhyan outcrop (Fig. 4). It exhibits a fining-upward trend. The sandstone of tidal bar origin occurs at the bottom part of the formation, and it gradationally passes upwards into a fining-upward shale–sandstone alternation. The shale becomes sand-free at the top, and organic carbon-rich, containing pyrite and locally phosphate (Banerjee et al. 2006a; Sur et al. 2006; Schieber et al. 2007). A basin-wide ash tuff of ~10 m in thickness demarcates the bottom of the Upper Kaimur. Locally, this tuff overlies the Lower Vindhyans (Chakraborty et al. 1996). The Upper Kaimur begins with a shelf storm succession (Chakraborty and Bose 1992) that gradationally gives way upwards into a coarsening-upward succession consisting of fluvial and eolian deposits. The overlying Rewa Shale at the lower part of the Rewa Formation consists of a storm-infested shelf deposit. A granular lag blanket occurs at the top of the Kaimur Formation, below the Rewa Formation. A diamondiferous conglomerate occurs at the base of the Rewa Shale. Chakraborty et al. (1996) reported a volcaniclastic deposit above the conglomerate. The Rewa Shale shelf succession depicts stacking of shoaling-upward facies couplets separated from each other by thin well-sorted granular lags. The Rewa Shale passes upwards into the Rewa Sandstone,

Table 2.18 Description and interpretation of constituent facies of the Upper Bhander Sandstone around Maihar area

Facies	Description	Interpretation
Facies A (Wedge-shaped, wave-imprinted sandstone)	It comprises vertically stacked, broadly wedge-shaped beds of sandstone grading into mudstone. Bed thickness decreases upwards in every such stack. The sandstone is the dominant component of the facies, moderately sorted, fine-grained, reddish in colour gradually passes upwards into silty mudstone. Mud clasts are present within the sandstone beds. Soles of the basal sandstone bear gutters, bipolar prod marks and locally flutes. Every sandstone bed is characterized by planar to broadly wavy, and hummocky laminae and wave ripple laminae on top with concomitant decline in grain size. Superimposed, interference and ladder ripples, polygonal cracks, rain imprints and warts are common on bed surface	Bed internal and surficial structures reflect the presence of strong oscillation components within the flows responsible for deposition of the sediments. The gradational change in lithology and structure from bottom to top of the bed indicate that these were deposited within a relatively short time during individual flow event. The flows evidently waned through time. Deposition from storm-generated combined, wave-cum-current, is thus inferred
Facies B (Mixed facies consists of three subfacies)		
Thinly stratified sheet–sandstone with wrinkle laminae	Composed of reddish, fine to medium-grained flat bedded sandstone having average thickness of 5.2 m. The sub-facies is composed of low-angle thin, planar laminae. They are laterally persistent for several metres. Individual laminae are characterized by concentration of coarsest grain fraction at the top. The sandstone is well sorted and fine-grained. Occasionally strongly asymmetric ripples having low amplitude, sharp, straight to sinuous crest and coarser grain concentration along the crest. Irregular crinkly laminated adhesion laminae are present in flat beds (less than a cm to 15 cm in thickness). Along with minute and wrinkled adhesion ripples with discontinuous crest line, warts are present on bed surface. Sand fill polygonal cracks are common on the bed surface	Structures present within the subfacies clearly indicate eolian origin. All these structures are typically found in eolian sand sheets and likely to be interdune deposits. The adhesion ripples and adhesion cross-laminae required high substrate moisture for efficient trapping of the sand grains that struck on the surface. Adhesion warts also formed by coagulation of deflated sand in the presence of moisture

(continued)

Table 2.18 (continued)

Facies	Description	Interpretation
Isolated medium-scale cross- stratified sandstone	Characterized by fine- to medium-grained well-sorted planar cross-stratified and lenticular sandstone bodies encased by earlier subfacies. The cross-sets are solitary in occurrence (ca. 60 cm thick), may have scalloped bases and show regular alterations of inversely graded grainflow and normally graded grainfall laminae. The cross-strata may have short asymptotic toes, laterally traceable few tens of metre	This solitary cross-sets internally characterized by regular grainflow and grainfall alternations and encased by eolian sand sheet deposits likely to be formed by migration of isolated eolian dunes
Channelized sandstone	Lenticular sandstone bodies characterized by concave-upward base and flat top with a channel fill geometry. The sandstones are fine- to medium-grained and less sorted. The sandstone lenses measure up to 3.6 m in length and 1.2 m in thickness. Mud clast is concentrated at the base of sandstone. The sandstone bodies consist of flat bedded sandstone of average thickness of 15 cm, vertically stacked up. The beds occasionally topped by thin layers of eolian ripples or translatent strata. The channel fill succession may have mudstone (thickness up to 10 cm) at the top. This mudstone often bears polygonal cracks. These channel fill bodies are encased by eolian sand sheet deposits. The sandstone beds may be planar or cross-stratified (set thickness up to 45 cm), reactivation surface present. Occasionally, sigmoidal cross-strata are present within the sandstone	The relatively poor sorting of the sediment and their channel fill geometry is likely result of deposition in streamlets. Thin mud partings between beds or thin lenses of eolian deposit, between and within the beds suggest that the streams were of ephemeral nature. Fluvial deposition was apparently initiated suddenly and did not last long. The presence of polygonal cracks of possible desiccation in these mudstones suggests emergence, draining out of water. Vertical upward transition from cross-strata to planar laminae without any significant change in grain size within a channel-filled succession indicates a transition from lower flow regime to upper flow regime. The transition is attributable to late-stage shit flood in shallow, broad ephemeral streams

(continued)

Table 2.18 (continued)

Facies	Description	Interpretation
Facies C (Large-scale cross-stratified sandstone)	This facies is composed of well-sorted, fine- to medium-grained sandstone characterized by simple or compound cross-bedding and reddish in colour. Cross-sets range in thickness from ca. 25 cm to 2.2 m. The cross-strata occur mostly in cosets. Planar laminae sets also intervene two cosets. The cross-beds often exhibit slump and penecontemporaneous deformation. Relatively coarser and finer foreset often alternate amongst them. The coarser foreset generally wedges downslope, while the finer foreset wedge upslope. The coarser foreset at places shows reverse grading, while the finer forests exhibit normal grading. Individual grainflow laminae are few mm to 7 cm thick	This facies likely to be a draa, made principally of dune deposits. Regular alternations between grainflow and grainfall strata and inverse grading in grainflow laminae are typical of eolian dunes

Sequence	Formation		Description	Paleogeography	Systems Tract
4500 (m) UPPER VINDHYAN	Bhander	Upper Bhander Sandstone	Well sorted sandstone with wave features below and cross-strata, translatent strata and adhesion laminae above (CU)	Fluvio-eolian and marginally marine	HST
		Sirbu Shale	Sandstone-shale alternations with wave and quadripolar sole features. Carbonate patches with emergence features below (CU)	Shelf / Lagoon	←— MFS
		Lower Bhander Sandstone	Sandstone-mudstone alternations, with wave and emergence features (NT)	Coastal	TST
		Bhander Limestone	Micritic, oolitic and stromatolitic limestone (NT)	Shallow marine	
		Ganurgarh Shale	Mudstone-sandstone interbeds, wave and emergence features (NT)	Chenier	
	Rewa	Rewa Sandstone	Well-sorted sandstone, bimodal cross-strata below and assorted sandstone, unimodal x-strata and emergence features above (CU)	Tidal to fluvio-eolian	HST
		Rewa Shale	Shale-sandstone alternations with wave and sole features (CU)	Shelf	←— MFS / TST
	Kaimur	Upper Kaimur	Well sorted sandstone with thin mudstone below and large x-stratified coarse sandstone with emergence features above (CU)	Shelf in fluvio-eolian	HST
		Lower Kaimur	Sandstone-shale interbeds with waves and sole features sand-free black shale above (FU)	Intertidal to shelf	←— MFS / TST
LOWER VINDHYAN	Rohtas	Rohtas limestone	Limestone, locally intraclastic, wave rippled (NT)	Shelf	HST ←— MFS
		Rampur Shale	Grey shale with sand-filled gutters (FU)	Shelf	TST
	Kheinjua	Chorhat Sandstone	Well sorted, wave features sandstone, often amalgamated (CU)	Shallow marine	HST
		Koldaha Shale	Shale-sandstone interbeds with wave and sole features. Local coarser poorly sorted sandstone intervals (CU)	Dominantly shelf, deltaic fluvial	←— MFS / TST
	Porcellanite		Volcanic ash, pyroclastic flow/surge deposits (NT)	Shallow marine	
	Kajhrahat	Kajhrahat Limestone	Dolomitized limestone with stromatolitic and desiccation features towards top (NT)	Subtidal to peritidal	HST ←— MFS
		Arangi Shale	Scarcely exposed grey shale	Shelf	
0 (m)	Deoland		Well sorted sandstone, bimodal-bipolar cross-stratification. Localized basal diamictile (FU)	Shallow shelf	TST

Fig. 2.38 Paleogeographies, sequences and systems tracts in the Vindhyan Supergroup (CU = coarsening-upward; FU = fining-upward; NT = no distinct trend)

which is upward coarsening. The sandstone exhibits shallow marine-tidal deposits grading upwards into fluvial/eolian deposits (Bose and Chakraborty 1994). The overlying Bhander Formation contains stromatolitic and oolitic patches and alternating grey shale at the lower part. This part has been interpreted as a lagoonal deposit by Singh (1973). An organic-rich, sand-free shale, devoid of emergence features overlies the lagoonal shale, locally intervened by a few cm-thick conglomerate bed. The shelfal Sirbu Shale exhibits an overall coarsening-upward trend, incorporates more storm originated siltstone and sandstone beds with abundant bipolar sole marks and gradationally passes upwards into the Upper Bhander Sandstone, which records the youngest Vindhyan deposit (Sarkar et al. 2002b). The latter consists of coastal storm bed packages, eolian dunes and fluvial deposits at certain levels (Bose et al. 1999; Sarkar et al. 2004).

Sedimentary rocks at the base of the Lower Kaimur, despite occurrence on top of an unconformity, are never coarser than granule grain size. The preferential occurrence of the Lower Kaimur in the southeastern sector of the basin points to differential subsidence in a landward direction and a sediment depocentre there. The Lower Kaimur is recognized as a TST with the MFS at its top (Bijaigarh Shale), because of its overall fining-upward trend and typical condensed zone sedimentary rocks at its top (Bose et al. 2001; Banerjee et al. 2006a). Sharp upward passage of the condensed zone to the shallow shelf sandstone at the base of the Upper Kaimur is a definite indicator of rapid regression. The shelf to terrestrial transition in the Upper Kaimur is progradational, but its interpretation in terms of sequence stratigraphy contains an inherent dichotomy. The Rewa shale on top of the Kaimur Formation, blanketed by a granular lag, indicates a major transgression. The MFS is, therefore, at the base of the Rewa Shale where the basal organic-rich, greenish-black shale that incorporates thin undiluted ash layers represents the condensed zone. The Rewa shale is generally aggradational, although it includes a few smaller progradational cycles. The Rewa Shale exhibits slow rise of relative sea level. The aggradational succession and the rapid progradational Rewa Sandstone together form the HST. On top of the fluvial/eolian Rewa Sandstone, the marginal marine Ganurgarh Shale, Bhander Limestone, Lower Bhander Sandstone and lagoonal Sirbu Shale together can be considered as aggradational in consequence of a slow rise in relative sea level. The sharp upward transition from the lagoon to the deep shelf within the Sirbu Shale succession speaks for a rapid transgression. Succeeding the progradational Rewa Sandstone, this aggradational succession is interpreted as a TST. The shelfal Sirbu Shale and the dominantly terrestrial/marginal marine Upper Bhander Sandstone together constitute an HST with the MFS at the base.

Coordinates of Locations Mentioned in this Chapter

Chopan	24°31′23.78″N; 83° 1′12.12″E
Sidhi	24°23′43.44″N; 81°52′58.09″E
Shikarganj	24°17′13.33″N; 81°27′48.86″E
Koldaha	24°24′27.7″N; 81°40′31.4″E
Dhanwahi	23°58′53.1″N; 80°48′06.6″E
Chorhat	24°25′33.90″N; 81°40′9.06″E
Rampur Naikin	24°20′32.48″N; 81°28′33.52″E

(continued)

(continued)

Mohaniya	24°26′24.96″N; 81°37′11.92″E
Kudari	24°7′52.87″N; 81°0′43.05″E
Ghurma	24°37′16.62″N; 83°5′42.80″E
Dala	24°27′8.97″N; 83°2′32.41″E
Churk	24°38′59.96″N; 83°5′59.98″E
Mirzapur	25°8′0.48″N; 82°33′51.83″E
Rewa	24°31′47.85″N; 81°17′59.80″E
Satna	24°36′0.49″N; 80°49′53.00″E
Maihar	24°16′9.89″N; 80°45′23.94″E
Katni	23°49′50.65″N; 80°24′25.91″E
Jadunathpur	24°31′27.7″N; 83°35′31.5″E
Bandu	24°32′53.96″N; 83°54′33.79″E
Amjhore	24°39′18.39″N; 83°58′40.82″E
Kalinjar	25°51′13.10″N; 82°28′28.89″E
Chitrakut	25°10′43.31″N; 80°51′55.67″E
Sasaram	24°57′13.97″N; 84°0′51.42″E
Hanumana	24°46′29.80″N; 82°5′24.00″E
Govindgarh	24°22′49.11″N; 81°18′0.04″E
Bhadanpur	24°9′35.47″N; 80°49′2.51″E
Sarlanagar	24°12′4.62″N; 80°48′6.56″E
Dolni	24°18′10.72″N; 80°48′11.49″E
Girgita	24°14′55.29″N; 80°48′26.44″E
Emilia	24°15′28.89″N; 80°48′18.29″E
Babupur	24°38′24.54″N; 80°55′39.67″E
Sajjanpur	24°33′12.94″N; 80°58′22.07″E
Sagmanhia	24°11′30.35″N; 80°47′36.36″E
Parasmania	24°22′7.35″N; 80°37′42.58″E

References

Ahmad F (1955) Glaciation in the Vindhyan system. Curr Sci 24:231

Ahmad F (1958) Paleogeography of Central India in the Vindhyan Period. Rec Geol Surv India 87:513–548

Altermann W, Herbig HG (1991) Tidal flat deposits of the lower Proterozoic Campbell Group along the southwestern margin of the Kaapvaal Craton, northern Cape Province, South Africa. J African Ear Sci 13:415–435

Arnott RWC (1993) Quasi-planar-laminated sandstone beds of the Lower Cretaceous Bootlegger Member, north-central Montana: evidence of combined flow sedimentation. J Sed Pet 63:488–494

Auden JB (1933) Vindhyan sedimentation in the Son valley, Mirzapur District. Mem Geol Surv India 62:141–250

Banerjee S (1997) Facets of Mesoproterozoic Semri sedimentation, Son valley, M.P. PhD Thesis, Jadavpur University, Kolkata, India

Banerjee S (2000) Climate versus tectonic control on the storm cyclicity in Mesoproterozoic Koldaha Shale, Vindhyan Supergroup, Central India. Gond Res 3:521–528

Banerjee S (2010) Distinction between marine and continental facies in Precambrian sedimentary succession: palaeoproterozoic Deoland Formation, Vindhyan Supergroup, central India. Gond Geol Mag 25:239–250

Banerjee S, Jeevankumar S (2003) Facies motif and paleogeography of Kheinjua Formation, Vindhyan Supergroup, eastern Son valley. Gond Geol Mag Spec Pub 7:363–370

Banerjee S, Jeevankumar S (2005) Microbially originated wrinkle structures on sandstone and their stratigraphic context: Paleoproterozoic Koldaha Shale, central India. Sed Geol 176:211–224

Banerjee S, Jeevankumar S (2007) Facies and depositional sequence of the Mesoproterozoic Rohtas Limestone: eastern Son valley, India. J Asian Earth Sci 30:82–92

Banerjee S, Sarkar S, Bhattacharyya SK (2005) Facies, dissolution seams and stable isotope characteristics of the Rohtas Limestone (Vindhyan Supergroup) in the Son valley area, central India. J Earth Sys Sci 114:87–96

Banerjee S, Dutta S, Paikaray S, Mann U (2006a) Stratigraphy, sedimentology and bulk organic geochemistry of black shales from the Proterozoic Vindhyan Supergroup (central India). J Earth Sys Sci 115:37–48

Banerjee S, Jeevankumar S, Sanyal P, Bhattacharyya SK (2006b) Stable isotope ratios and nodular limestone of the Proterozoic Rohtas Limestone: Vindhyan basin, India. Carbonates Evaporites 21:133–143

Banerjee S, Bhattacharya SK, Sarkar S (2007) Carbon and oxygen isotopic variations in peritidal stromatolite cycles, Paleoproterozoic Kajrahat Limestone, Vindhyan basin of central India. J Asian Earth Sci 29:823–831

Banerjee S, Jeevankumar S, Eriksson PG (2008) Mg–rich ferric illite in marine transgressive and highstand systems tracts: examples from the Paleoproterozoic Semri Group, central India. Precam Res 162:212–226

Basu A, Bickford ME (2015) An alternate perspective on the opening and closing of the intracratonic Purana basins in peninsular India. J Geol Soc India 85:5–25

Beukes NJ, Lowe BR (1989) Environmental control on diverse stromatolite morphologies in the 3000 Myr Pongola Supergroup. Sedimentol 36:383–397

Bhattacharyya A (1996) (ed) Recent Advances in Vindhyan Geology. Mem Geol Soc India 36:1–331

Bhattacharyya A, Morad S (1993) Proterozoic braided ephemeral fluvial deposits: an example from the Dhandraul Sandstone Formation of the Kaimur Group, Son valley, central India. Sed Geol 84:101–114

Bickford ME, Mishra M, Muelle PA, Kamenov GD, Schieber J, Basu A (2017) U-Pb age and Hf isotope compositions of magmatic zircons from a rhyolite flow in the Porcellanite Formation in the Vindhyan Supergroup, Son Valley (India): implications for its tectonic significance. J Geol 125:367–379

Bose PK, Chakraborty PP (1994) Marine to fluvial transition: Proterozoic Upper Rewa Sandstone, Maihar, India. Sed Geol 89:285–302

Bose PK, Chaudhuri AK, Seth A (1988) Facies, flow and bedform patterns across a storm-dominated inner continental shelf: Proterozoic Kaimur Formation, Rajasthan, India. Sed Geol 59:275–293

Bose PK, Banerjee S, Sarkar S (1997) Slope-controlled seismic deformation and tectonic framework of deposition of Koldaha Shale, India. Tectonophy 269:151–169

Bose PK, Chakraborty S, Sarkar S (1999) Recognition of ancient eolian longitudinal dune: a case study in Upper Bhander Sandstone, Son valley, India. J Sed Res 69:86–95

Bose PK, Sarkar S, Chakraborty S, Banerjee S (2001) Overview of the Meso-to Neoproterozoic evolution of the Vindhyan basin, Central India. Sed Geol 141:395–419

Bose PK, Sarkar S, Das NG, Banerjee S, Mandal A, Chakraborty N (2015) Proterozoic Vindhyan basin: configuration and evolution. Mem Geol Soc London 43:85–102

Cas RAF, Wright JV (1987) Volcanic Successions. Allen and Unwin, London, Boston, Sydney, Wellington

Catuneanu O (2006) Principles of Sequence Stratigraphy. Elsevier, Amsterdam

Chakraborty C (1993) Morphology, internal structure and mechanics of small longitudinal (seif) dunes in an aeolian horizon of Proterozoic Dhandraul Quartzite, India. Sedimentol 40:79–85

Chakraborty C (1994) Proterozoic Kaimur Formation, Son valley, India: Facies and sequence in tectonogeographic frame with special bearing on mechanics of clastic sedimentation. Unpubl PhD Thesis, Jadavpur University, Kolkata

Chakraborty C (1995) Gutter casts from the Proterozoic Bijaigarh Shale Formation, India: their implications for storm-induced circulation in shelf settings. Geol J 30:69–78

Chakraborty C (1996) Sedimentary records of erg development over a braid plain: Proterozoic Dhandraul Sandstone, Vindhyan Supergroup, Son Valley. In: Bhattacharyya A (ed) Recent Advances in Vindhyan Geology. Mem Geol Soc India 36, pp 77–99

Chakraborty C, Bose PK (1992) Rhythmic shelf storm beds: Proterozoic Kaimur Formation, India. Sed Geol 77:259–268

Chakraborty PP, Banerjee S, Das NG, Sarkar S, Bose PK (1996) Volcaniclastics and their sedimentological bearing in Proterozoic Kaimur and Rewa Groups in Central India. In: Bhattacharyya A (ed) Recent Advances in Vindhyan Geology, Mem Geol Soc India 36, pp 1011–1126

Chakraborty PP, Sarkar S (2005) Episodic emergence of offshore shale and its implication: Late Proterozoic Rewa Shale, Son Valley, central India. J Geol Soc India 66:699–712

Chakraborty PP, Sarkar S, Bose PK (1998) A viewpoint on intracratonic chenier evolution: clue from a reappraisal of the Proterozoic Ganurgarh Shale, central India. In: Palliwal BS (ed) The Indian Precambrians. Scientific Publishers, Jodhpur, pp 61–72

Chakraborty T, Sarkar S, Chaudhuri A, Dasgupta S (1996) Depositional environments of Vindhyan and other Purana basins: a reappraisal in the light of recent findings. In: Bhattacharyya A (ed) Recent Advances in Vindhyan Geology, Mem Geol Soc India 36, pp 101–126

Chaudhary MS (1953) Late Precambrian glaciation and its relationship with the Bijawar Series in the type are of Bundelkhand. In: Proceedings 40th. Ind Sci Congr vol 3, p 20

Dott RH Jr, Bourgeois J (1982) Hummocky cross stratification: significance of its variable bedding sequences. Geol Soc Am Bull 93:663–680

Dubey BS, Chaudhary MS (1952) Late Precambrian glaciation in Central India. Curr Sci 21:331–332

Galloway WE (1990) Paleogene depositional episodes, genetic stratigraphic sequences and sediment accumulation rates, NW Gulf of Mexico basin. GCSSEPM Foundation, Eleventh Annual Research Conference, Programs and abstracts, pp 165–176

Ghosh SK (1971) Petrology of Porcellanite rocks of the Samaria area, Sidhi District, Madhya Pradesh. Quart J Geol Min Met Soc India 43:153–164

Glumac B, Walker KR (1997) Selective dolomitisation of Cambrian microbial carbonate deposits: a key to mechanisms and environments of origin. Palaios 12:98–110

Gomez JJ, Fernandez-Lopez S (1994) Condensation processes in shallow platforms. Sed Geol 92:147–159

Grotzinger JP (1986) Cyclicity and paleoenvironmental dynamics, Rocknest platform, northwest Canada. Geol Soc Am Bull 97:1208–1231

Grotzinger JP (1989) Facies and evolution of Precambrian carbonate depositional systems: emergence of the modern platform archetype. In: Crevello PD, Wilson JL, Sarg JF, Read JF (eds) Controls on Carbonate Platform and Basin Development. SEPM Spec Publ 44, pp 79–106

Kidwell SM (1991) Condensed deposits in siliciclastic sequences: expected and observed features. In: Einsele G, Ricken W, Seilacher A (eds) Cycles and Events in Stratigraphy. Springer-Verlag, Berlin, pp 682–695

Lajoie J, Stix J (1992) Volcaniclastic rocks. In Walker R, James N (ed) Facies Models. Geol Ass Can, pp 101–118

Lowe DR (1982) Sediment gravity flows: II depositional models with special reference to the deposits of high–density turbidity currents. J Sed Petrol 52:279–297

Mathur SM (1954) Late Precambrian glaciation in Central India-rejoinder. Curr Sci 23:7–8

Mathur SM (1960) A note on the Bijawar Series in the eastern part of the type area. Rec Geol Surv India 86:539–544

Mathur SM (1981) The middle Proterozoic Gangau tillite, Bijawar Group, Central India. In: Hambrey MJ, Harland WB (eds) Earth's Pre-Pleistocene Glacial Records. Cambridge University Press, Cambridge, pp 424–427

Mishra M, Bickford ME, Basu A (2018) U-Pb Age and Chemical Composition of an Ash Bed in the Chopan Porcellanite Formation, Vindhyan Supergroup, India. J Geol 126:553–560

Orton GJ (1996) Volcanic environments. In: Reading HG (ed) Sedimentary Environments: Processes, Facies and Stratigraphy. Blackwell, Oxford, pp 485–569

Pflüger F, Seilacher A (1991) Flash flood conglomerates. In: Einsele G, Ricken W, Seilacher A (eds) Cycles and Events in Stratigraphy. Springer, Berlin, pp 383–391

Roy S, Banerjee S (2002) Facies and petrography of the Porcellanite Formation around Chopan, Uttar Pradesh. J Indian Ass Sedimentol 20:195–205

Samanta P, Mukhopadhyay S, Eriksson PG (2016) Forced regressive wedge in the Mesoproterozoic Koldaha Shale, Vindhyan basin, Son valley, central India. Mar Petrol Geol 71:329–343

Sangree JB, Vail PR, Mithcum R Jr (1990) A summary of exploration applications of sequence stratigraphy. GCSSEPM Foundation, Eleventh Annual Research Conference, Program and abstracts, pp 321–327

Sarg JF (1988) Carbonate sequence stratigraphy. In: Wilgus C, Hastings B, Ross C, Posamentier H, Van Wagoner J, Kendall CGStC (eds) Sea Level Changes: An Integrated Approach. SEPM Spec Publ 42, pp 155–181

Sarkar S, Bose PK (1992) Variations in Late Proterozoic stromatolites over a transition from basin plain to nearshore subtidal zone. Precam Res 56:139–157

Sarkar S, Banerjee S, Bose PK (1996a) Trace fossils in the Mesoproterozoic Koldaha Shale, Central India, and their implications. N Jb Geol Paleo Mh 7:425–438

Sarkar S, Chakraborty PP, Bose PK (1996b) Proterozoic Lakheri Limestone, central India: facies, paleogeography, and paleophysiography. Geol Soc India Mem 36:5–26

Sarkar S, Banerjee S, Chakraborty C, Bose PK (2002) Shelf storm flow dynamics: insight from the Mesoproterozoic Rampur Shale, central India. Sed Geol 147:89–104

Sarkar S, Chakraborty S, Banerjee S, Bose, PK (2002b) Facies sequence and cryptic imprint of sag tectonics in late Proterozoic Sirbu Shale, central India. In: Altermann W, Corcoran P (eds) Precambrian Sedimentary Environments: a Modern Approach to Ancient Depositional Systems. Spec Publ IAS 33, Blackwell Science, pp 369–382

Sarkar S, Eriksson PG, Chakraborty S (2004) Epeiric sea formation on Neoproterozoic supercontinent break-up: a distinctive signature in coastal storm bed amalgamation. Gond Res 7:313–322

Sarkar S, Banerjee S, Samanta P, Jeevankumar S (2006) Microbial mat–induced sedimentary structures and their implications: examples from Chorhat Sandstone, M.P. India J Earth Sys Sci 115:49–60

Schieber J, Sur S, Banerjee S (2007) Benthic microbial mats in black shale units from the Vindhyan Supergroup, Middle Proterozoic of India: the challenges of recording the genuine article. In: Schieber J, Bose PK, Eriksson PG, Banerjee S, Sarkar S, Catuneanu O and Altermann W (eds) An Atlas of Microbial Mat Features Preserved within the Clastic Rock Record. Elsevier, pp 225–232

Schmincke HU, Van Den Boggard P (1991) Tephra layers and tephra events. In: Einsele G, Ricken W, Seilacher A (eds) Cycles and Events in Stratigraphy. Springer, Berlin, Heidelberg, New York, pp 392–429

Seilacher A, Bose PK, Pflüger F (1998) Triploblastic animals more than 1 billion years ago: Trace fossil evidence from India. Science 282:80–83

Singh IB (1973) Depositional environment of the Vindhyan sediments in the Son valley area. Recent Researches in Geology, vol 1. Hindusthan Publ Corp, New Delhi, pp 140–152

Singh AK, Chakraborty PP, Sarkar S (2018) Redox structure of Vindhyan hydrosphere: clues from total organic carbon, transition metal (Mo, Cr) concentrations and stable isotope (delta C-13) chemistry. Curr Sci 115:1334–1341

Sonnenfield MD (1996) Sequence evolution and hierarchy within the lower Mississippian Madison Limestone of Wyoming. In: Longman MW, Sonnenfield MD (eds) Paleozoic Systems of the Rocky Mountain Region. SEPM Rocky Mountain Section, Tulsa, Oklahoma, pp 1–28

Southgate PN (1989) Relationships between cyclicity and stromatolite form in the Late Proterozoic Bitter Springs Formation, Australia. Sedimentol 36:323–339

Srivastava RN (1977) Environmental significance of some depositional structures in banded porcellanites (Lower Vindhyan) of Mirzapur District, U.P. J Indian Ass Sediment 1:45–51

Sur S, Schieber S, Banerjee S (2006) Petrographic observations suggestive of microbial mats from Rampur Shale and Bijaigarh Shale, Vindhyan basin, India. J Earth Sys Sci 115:61–66

Van Wagoner JC, Mitchum RM, Campion KM, Rahmanian VD (1990) Siliciclastic sequence stratigraphy in well logs, cores and outcrops: concepts for high resolution correlation of time and facies. Methods in Exploration Series, Am Ass Petrol Geol 7:1–55

Walker RG, Plint AG (1992) Wave and storm dominated shallow marine system. In: Walker RG, James NP (eds) Facies Models, Response to Sea Level Change. Geol Ass Canada, Geotext 1, pp 9–238

Williams GE, Schmidt PW (1996) Origin and paleomagnetism of the Mesoproterozoic Gangau tilloid (basal, central India. Precam Res 79:307–325

Chapter 3
Selected Traverses

This chapter deals with selected traverses through the outcrops of Vindhyan Supergroup. Each traverse includes several stops to visit the nearby outcrops. Each traverse needs at least one full day to visit the outcrops and quickly examine the broad characteristics of each of the constituent formations. These traverses are planned from west to east, starting from Maihar, covering all formations comprising the Vindhyan Supergroup. Four of the seven traverses can be covered stationed at Maihar. Coordinates of all the places are provided at the end of this chapter.

3.1 Dhanwahi–Bhadanpur

A north–south traverse along the metalled road connecting Bhadanpur to Dhanwahi covers excellent outcrops of Kajrahat, Porcellanite, Kheinjua and Rohtas Formations (Fig. 3.1). The basal stratigraphic unit, the Deoland Formation, cannot be examined in this traverse.

Stop 1 Dhanwahi and Kuteswar Mines

The Kajrahat Limestone crops out on both northern and southern flanks of the Mahanadi River. It is also nicely exposed within the Kuteswar Limestone Mines (A1 in Fig. 3.1). The landscape of the area has been changed substantially because of the construction of Bansagar Dam in the year 2006, which has caused large-scale submergence of excellent outcrops of stromatolitic limestone under water. Table 2.2 provides a brief description of the constituent facies of the Kajrahat Formation in this area. A buff-coloured, massive dolostone interspersed with isolated planar-curved cross-stratified dolostone lenses occupies the basal division of the section. The middle division of the Kajrahat Limestone consists of vertical alternations of a dark-coloured, faintly laminated limestone and a buff-coloured dolostone with gypsum pseudomorphs at places. The upper part of the Kajrahat Limestone consists exclusively of stromatolitic limestone.

© Springer Nature Singapore Pte Ltd. 2020
S. Sarkar and S. Banerjee, *A Synthesis of Depositional Sequence of the Proterozoic Vindhyan Supergroup in Son Valley*, Springer Geology,
https://doi.org/10.1007/978-981-32-9551-3_3

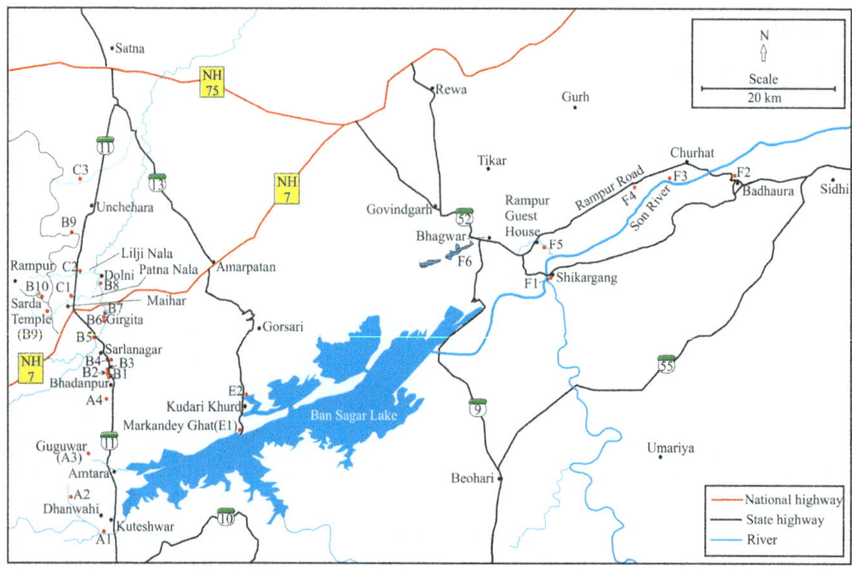

Fig. 3.1 Map showing traverses and field spots in and around Maihar (position of field spots and localities are marked by red and black dots, respectively)

The Kuteswar Mine section exposes the lower two divisions, while the upper division of the Kajrahat Limestone exhibits superb preservation of stromatolites, north of the River Mahanadi. The organo-sedimentary structures occur in three distinct size and shape groups (Fig. 3.2a–c). The average column height and head diameter of large stromatolites are 20 cm and 6 cm, respectively. The smaller stromatolites have average column height and head diameter 3.5 cm and 1.8 cm, respectively. The columns in the latter are branched, while it is not so in the former. The columns in the former are generally of uniform width and cone-shaped, but in the latter they are round-headed. Wavy–crinkly, low-relief microbial laminites form the third group of organo-sedimentary structures. V-shaped cracks are common in vertical sections. The organo-sedimentary structures occur in repeated cycles, with larger stromatolites at their base and microbial laminites at the top (Fig. 3.2b).

The Kajrahat Limestone Formation has been interpreted as shallow marine deposits (Bose et al. 2001; Banerjee et al. 2007). Occasional extreme shallowing is indicated by upright gypsum pseudomorphs in the dolostones and cracks of possible desiccation origin. The conical larger and unbranched stromatolites are common without any preferred orientation, while the smaller stromatolites exhibit strong imbrication. The former possibly formed in a relatively deeper part of the shelf. The cycles represent shoaling-up parasequences bounded below and above by marine flooding surfaces.

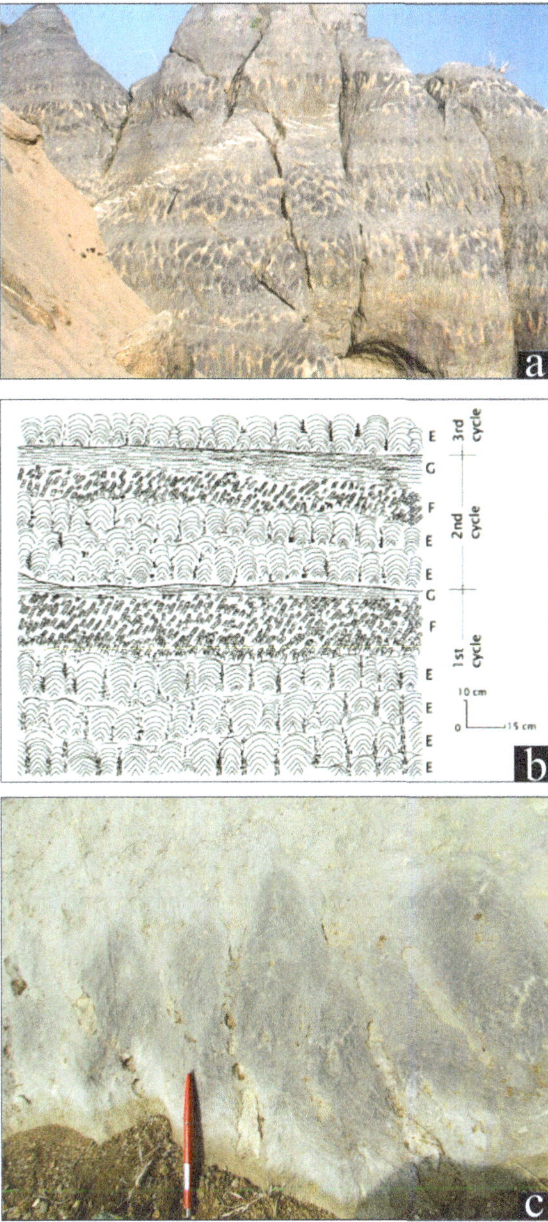

Fig. 3.2 Cyclic alternations between large- and small-scale stromatolites within the Kajrahat Limestone (**a**), sketch showing vertical facies transition forming cycles, each starting with large-scale stromatolites followed upwards by small-scale stromatolites and microbial laminite, (**b**) large stromatolite columns with prominent intercolumnar area (**c**) (hammer length in a = 38 cm, pen length in c = 14 cm)

Stop 2 Saraswati River Section

Banks of the River Saraswati, northwest of Dhanwahi, expose the Koldaha Shale of the Kheinjua Formation (A2 in Fig. 3.1). The colour of the shale varies from grey to black. The Koldaha Shale exhibits repeated alternations between shale and fine sandstone (Fig. 3.3). Sole features of various types, viz. gutter casts, prod marks, flutes occur below the sandstone beds. A storm-dominated, outer shelf origin has been suggested for the deposit (Banerjee 2007).

Stop 3 Guguwar Village

A low-relief hill near Guguwar exposes rippled sandstones of the Chorhat Sandstone. Dense vegetation has covered most outcrops in this spot (A3 in Fig. 3.1).

Stop 4 Limestone quarries south of Bhadanpur

The Rohtas Limestone, the uppermost member of the Semri Group is well exposed in several quarries, a few km southwest of Sarlanagar. Many of the quarries are located on the western side of SH11 (A4 in Fig. 3.4). Most of the facies (Sect. 3.2.6) constituting the Rohtas limestone are present in this area. A huge section of the Rohtas limestone is also exposed in the mine of Maihar Cement. However, it is difficult to visit exposures inside the mine because of the hectic movement of heavy vehicles. The Rohtas Limestone is constituted by two major facies association (Banerjee 1997; Banerjee et al. 2005, 2006a; Banerjee and Jeevankumar 2007). The primary structures in the lower facies association indicate deposition in a quiet environment (Fig. 3.4a–b). The upper facies association frequently exhibits cross-stratification, planar laminae, hummocky cross-stratification and wave ripples (Fig. 3.4c–f). Besides, graded bedding and flat pebble conglomerates are also very common in this part. All these structures indicate a high-energy depositional condi-

Fig. 3.3 Photograph showing outcrop of the Koldaha Shale

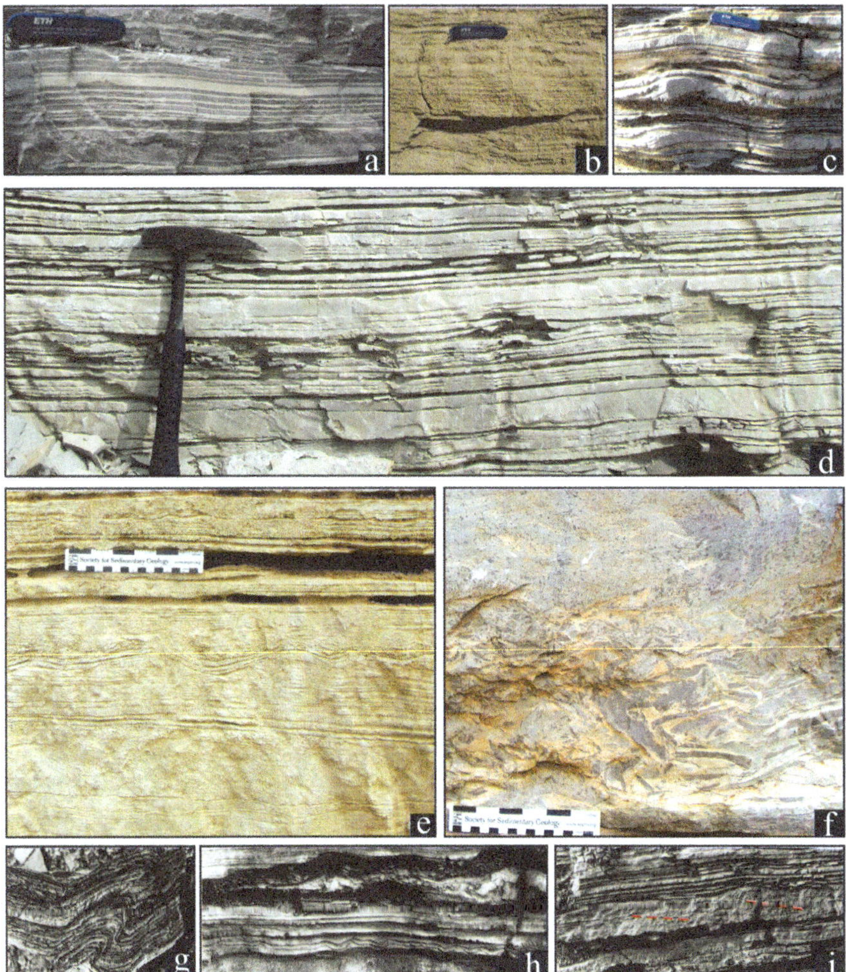

Fig. 3.4 Various primary and deformational features of the Rohtas Limestone: Alternations between dark- and light-coloured bands (**a**), crudely developed lamination (**b**) ripple lamination (**c**), planar lamination (**d**), transition from massive/graded bed to plane lamination, followed by ripple lamination (**e**), flat pebble conglomerate (**f**), minute fold and thrust (**g**), tectonic ripple (**h**), low-angle intra-stratal thrust (**i**) (Swiss knife length in a–c and h = 9.1 cm, hammer length = 38 cm, matchstick length in g and i = 4.5 cm)

tion. The detailed facies study reveals a distally steepened ramp depositional setting for the Rohtas Limestone (Banerjee 1997; Bose et al. 2001).

Asymmetric and locally recumbent folds occur between undisturbed beds (Fig. 3.4g). Locally the folds are top-truncated. Tectonic ripple, with sharp crests and troughs, is a ubiquitous feature of the planar-laminated limestone beds (Fig. 3.4h). These ripples exhibit micro-thrusts locally. Bed-confined sigmoidal thrusts occur in many places (Fig. 3.4i). The syn-depositional deformation structures occur at selected levels. The occurrence of deformation at the top of the Rohtas Limestone indicates a change in the tectonic setting of the Vindhyan from extensional to the compressional regime (Bose et al. 2001).

3.2 Bhadanpur–Maihar–Rampur

This traverse covers almost all the crucial sections of the Upper Vindhyan around Maihar area. The traverse begins with the oldest exposures of the Upper Vindhyan and ends with the top part of the Vindhyan Supergroup in the Maihar area.

Stop 1 Kaimur Ghat

This field spot is a road-cut through the Kaimur Hill (locally known as Kaimur Ghat) along the SH 11 joining Bhadanpur and Maihar. Rocks belonging to Kaimur and Rewa Formations are exposed in this section (B1 in Fig. 3.1). The uppermost Kaimur Sandstone exposed in this section has a thickness of ca. 20 m (Fig. 3.5a). The direct contact between the Rohtas Limestone and the Kaimur Sandstone is not exposed in this section. A porcelaneous shale/siltstone interval occur below the Kaimur Formation, although not at the direct contact. The sandstone is well sorted, medium- to fine-grained and is cross-stratified (Fig. 3.5b). A thin granular lag occurs immediately above this sandstone in the entire area (Fig. 3.5c). Small quartz pebbles are the major constituent of the lag deposit occurring above the Kaimur Sandstone. A grey shale, namely, Rewa Shale Member of the Rewa Formation overlies this granular lag. Sand interbeds common at the lower part of the shale become thinner towards the upper part. These sands are either massive- or planar-laminated. Flutes and different types of sole features are present at the soles of these sandstones.

Stop 2 Sandstone Quarry

This stop is located on a sandstone quarry (locally known as Khadan) about 500 m north of stop 1 towards Sarlanagar, on the western side of the road (B2 in Fig. 3.1). The lower marginal marine part of the Upper Rewa Sandstone, exposed in this quarry, is represented by fine-grained, very well-sorted sheet sandstones with abundant mud clasts at the base. The dominant palaeocurrent is directed towards the west. Large-scale cross-stratification and planar lamination are common primary structures within this well-sorted sandstone (Fig. 3.6a). The unidirectional, large-scale (set thickness varying from 80 cm to 2 m) cross-stratifications change style laterally in a rhythmic fashion, from (sigmoidal) convex- to concave-up. The planar-laminated units, varying in thickness from 60 to 90 cm, alternate with large-scale cross-stratified sandstones. The bedding plane surface is carved with parting lineations and current crescents (Fig. 3.6b). Orientations of these bed surface structures are the same as those present in cross-stratified units. The unidirectional large-scale cross-stratified unit form by the migration of sand waves. The rhythmic changes in the cross-stratification pattern closely resemble a tidal sequence (cf. Kreisa and Moila 1986). The planar-laminated

Fig. 3.5 Road cutting section showing outcrop of the Upper Kaimur Sandstone (**a**), Cross-stratified sandstone within the Upper Kaimur Sandstone (**b**), granular lag preserved on the bed surface at the transition between Upper Kaimur Sandstone and Rewa Shale (**c**)

sandstone in close association, decorated with parting lineations and current crescents, closely resembles a high energy beach (Bose and Chakraborty 1994).

Stop 3 Saralanagar Sections

This stop is located just after crossing the conveyer belt of the Maihar Cement towards Saralanagar and is about 100 m north of stop 2 (B3 in Fig. 3.1). The Upper Rewa Sandstone is nicely exposed on the western side of SH 11. The sandstones of this part of the Upper Rewa are fine- to coarse-grained, even granule-rich. Some contain rip-up mud clasts. Textural characteristics of these rocks differ from those in the lower marginal marine association of the Upper Rewa Sandstone. The sandstone is

Fig. 3.6 Cross-stratification (**a**) and current crescent (**b**) within the basal part of the Upper Rewa Sandstone of marine origin

strikingly less sorted and is thoroughly cross-stratified in this section. The paleocurrent direction measured from these cross-stratification spans over 180°, with mean towards the northwest. Locally eolian sandstone is present within an overall braided fluvial deposit. The fluvial association consists of a number of facies, with different scales and arrangements of cross-stratification as well as grain size.

Stop 4 Hathi Nala

This stop is close to the Sarlanagar bus stand, from where one needs to approach Hathi Nala (B4 in Fig. 3.1). It is the best section exposing the top part of the Upper Rewa Sandstone Member around Maihar. Several constituent facies of the Upper Rewa Sandstone crop out in this area. The granule-rich sandstone is the coarsest amongst all varieties of sandstone. The granule content occasionally exceeds 30% at the bases of the cross-stratified sandstones (Fig. 3.7a). The average cross-set thickness of the lensoid granular-rich bodies is ca.15 cm. Planar cross-stratified granular sandstone is another variety of sandstone (Fig. 3.7b). This variety is characterized by planar cross-stratified, solitary lensoid bodies of thickness up to 80–90 cm overlying planar or broadly undulating master erosion surfaces. Granules generally demarcate the foresets bases. The trough cross-stratified sandstone is the characteristic facies of the Upper Rewa Sandstone. This sandstone is granule-free, medium-grained, trough cross-stratified, forming lenticular bodies. The cross-set thickness is ca. 68 cm. The down-dipping cross-stratified sandstone is another commonly occurring facies. This is characterized by compound cross-stratification (Banks 1973) with major cross-strata dipping at ca 3°, and the smaller internal planar cross-laminae makes an angle of about 16° with them, in the same direction. They also show broad lensoid geometry. Reactivation surface, overturned cross-stratification and slump structure are frequently present within the Hathi Nala section (Fig. 3.8a, b).

Sandstones of this association having a low degree of textural maturity are not likely products of tide or wind. The vertical and lateral continuity with the mixed facies association, bearing eolian imprints indicates a fluvial origin. The Proterozoic River usually shows a braided character in the absence of vegetation and overbank

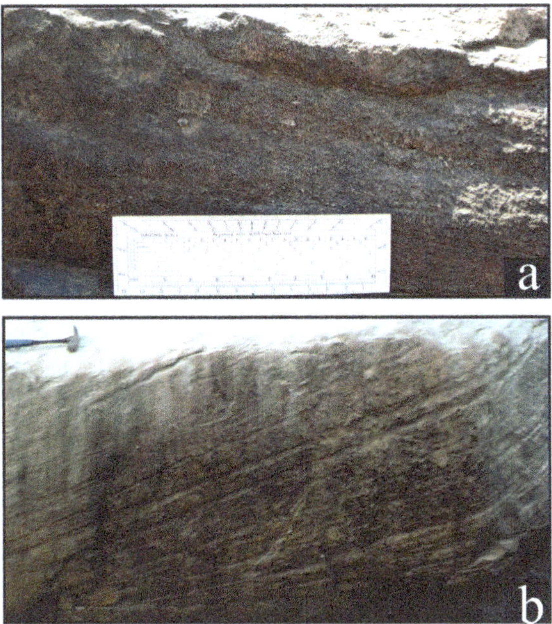

Fig. 3.7 Cross-stratified, granule-rich sandstone (**a**) and large-scale planar cross-stratification within the Upper Rewa Sandstone (**b**) (hammer length = 38 cm)

Fig. 3.8 Reactivation surfaces (marked by arrows) within cross-sets (**a**) and overturned cross-stratified sandstone (**b**) within the Upper Rewa Sandstone (hammer length = 38 cm)

mud (Schumm 1968; Cotter 1978; Fuller 1985). Stacks of extensive tabular or broad lensoidal sandbodies, bounded by planar or broadly undulated master erosion surfaces, are likely to be the products of a sandy braided river system (Long 1978; his Fig. 22). Scarcity of channel forms, fining-up short sequences and the complete absence of muddy overbank deposits are consistent with this idea. The slump folds represent a rapid shifting of the river channels undercutting the unstable banks.

Stop 5 Thomas River

This field spot is immediately after crossing the bridge over Thomas river (B5 in Fig. 3.1). The riverbank exposes the red-coloured Ganurgarh Shale Member of the Bhander Formation. The direct contact between the underlying Rewa Formation and the Ganurgarh Shale is not exposed in this area. The sandstone/siltstone horizons alternate with mudstone (Fig. 3.9). The constituent beds dip 3–5° towards northwestwards. Both wave and current ripples mantle the bed surfaces. Desiccation cracks are also abundant on bed surfaces. The gradational contact between the Ganurgarh Shale and the overlying the Bhander Limestone is exposed in this section. Lithological association, colour and primary sedimentary structure within the Ganurgarh Shale suggest deposition along a shoreline undergoing intermittent exposure. However, the Ganurgarh Shale also represents chenier plain in places (Chakraborty et al. 1998).

Stop 6 Girgita Mines

The access for this field spot deviates from SH 11 as an unmetalled road towards Girgita Mines. The Bhander Limestone is exposed in several quarries (B6 in Fig. 3.1). Most of the facies constituting the Bhander Limestone are exposed in this area. Limestone varieties include cross-stratified, plane-laminated, ripple-laminated, brecciated and stromatolitic. Stromatolite mounds are well exposed occasionally (Fig. 3.10a–g). Stromatolite columns may be inclined in places. Molar tooth structures are abundant at selected intervals (Fig. 3.10g). The detailed facies analysis reveals the deposition of the Bhander Limestone in a shallow shelf.

Fig. 3.9 Sandstone–siltstone alternation within the Ganurgarh Shale

Fig. 3.10 Photographs showing various features of the Bhander Limestone: Ripple laminae (**a**), plane lamination (**b**), cross-stratification (**c**), brecciation (**d**), stromatolite mound (**e**), inclined columns (**f**), molar tooth structure (**g**) (pen length = 14 cm, hammer length = 38 cm)

Stop 7 Amiliya Village

This stop is located on the banks of the Thomas River, exposing the Bhander Limestone near Amiliya village (B7 in Fig. 3.1). Facies constituting the Bhander Limestone in this area is discussed in Sect. 2.9.2. Planar-laminated and stromatolite facies dominate in this area. Stromatolites vary widely in size and shape (Fig. 3.11a–e). Inclined stromatolites are particularly well preserved in this area (Fig. 3.11e).

Stop 8 Dolni Village

This stop is located along the Thomas river beds near the Dolni village (B8 in Fig. 3.1). Several horizons in this section exhibit syn-depositional deformation features (see

Fig. 3.11 Stromatolites variability within the Bhander Limestone: vertical columns (**a**), branching of columns (**b** and **c**), irregular columns (**d**) and inclined columns (**e**) (pen length = 14 cm)

Sarkar et al. 2014a for details). Convolute laminae are present within a few limestone beds (Fig. 3.12a). The thickness of the convolute-laminated beds varies to a great extent. Besides, cross-stratified, ripple-laminated and stromatolite facies are also well exposed in this section (Fig. 3.12b).

Stop 9 Sarda Temple

The Sirbu Shale Member is nicely exposed along the hill range, which includes Sarda temple (B9 in Fig. 3.1). The entire hill range in this area nicely exposes the Sirbu Shale. The shale forms steep slopes, while the tops of the hill range are capped by the Upper Bhander Sandstone, the topmost member of the Bhander Formation. The Sirbu Shale is entirely terrigenous, except for a few oolitic and stromatolitic carbonate patches at its base (see Sect. 2.9.4). Beds are subhorizontal and broadly warped locally. The colour of the shale varies from grey to dark grey, even black at places (Fig. 3.13). The Sirbu Shale consists of alternations between mudstone, siltstone and sandstone. At the lower part of the hill range, the shale is almost devoid of sand, possibly representing the deeper part of the basin. Most of the siltstone and sandstone beds at the mid-level of Sirbu Shale are laterally persistent and exhibit gutter casts. The lower contacts of these layers are always sharp compared to its upper contact. The undersurface of the sandstone beds bears tool marks. The sand parting

Fig. 3.12 Convolute lamination (**a**) and small-scale stromatolites (**b**) within the Bhander Limestone (hammer length = 38 cm, pen length = 14 cm)

Fig. 3.13 Black-coloured shale within the Sirbu Shale near the Sarda Temple (hammer length = 14 cm)

gradually increases upwards. The facies association of the Sirbu Shale indicates deposition in a storm-dominated siliciclastic shelf (Sarkar et al. 2002a, 2005). The gradual upward transition from the Sirbu Shale to the Upper Bhander Sandstone indicates a progradation of the sea.

Stop 10 Rampur Road Section

The gradational contact between the Sirbu Shale and the Upper Bhander Sandstone is nicely exposed along a road from the Sarda Temple towards the Rampur village through the hill (B10 in Fig. 3.1). The section becomes sandier at the upper part. The sandstone is composed solely of stacked, broadly lenticular beds. The soles of the beds bear bipolar prod marks and locally flute casts (Fig. 3.14a). Lithology and internal structures vary systematically in bed-normal sections. Moderately sorted, fine-grained reddish sandstone grades upwards into brown mudstone. Sandstone beds are characterized by planar to broadly wavy and hummocky laminae passing upwards to wave-rippled laminae (cf. de Raaf et al. 1977). Besides, convolute-laminated beds are also common (Fig. 3.14b). At the sandstone-to-mudstone transition, the wave ripples are flat-crested and exhibit bidirectional cross-crest sand spillover (cf. Seilacher 1982). They have straight and locally bifurcated crests. Superimposed, interference and ladder ripples, desiccation cracks, rain prints and warts are common. Water recession marks are present locally.

A detailed facies analysis of the Upper Bhander Sandstone indicates deposition predominantly within a supralittoral zone (Sarkar et al. 2011). The vertical change in lithology and structure, without any sign of discontinuity, implies episodic deposition and waning and oscillatory flows, plausibly storm-induced (Bose et al. 1999). The sandstone present at the top part of the Upper Bhander Sandstone consists almost entirely of cross-stratified, well-sorted and fine-grained sandstones. The cross-sets and cosets range in thickness up to 2 m and 6.6 m, respectively. The foresets are planar with or without slightly asymptotic toes (Fig. 3.14c). Both inversely graded grainflow and normally graded grainfall laminae are present within the foresets. Thin

Fig. 3.14 Flute casts preserved under the sandstone bed (**a**) convolute-laminated sandstone bed (**b**) and cross-stratification with asymptotic toes (**c**) within the Upper Bhander Sandstone (hammer length = 38 cm, Swiss knife length = 9.1 cm)

(ca. 22 cm thick) sandstone beds characterized internally by eolian translatent strata or ripple laminae. This part of the red-coloured sandstone represents a draa, made up principally of dune deposits, locally interbedded with the thin interdune deposit (Bose et al. 1999). Considering the lithology and primary sedimentary structures, the sandstone represents a coastal setting with supralittoral storm, eolian sand sheet, draa, pond and ephemeral stream deposits (Bose et al. 1999; Sarkar et al. 2004a, 2011).

3.3 Maihar–Unchehara–Beta Traverse

This traverse covers mostly the Upper Vindhyan succession, north of Maihar. Most exposures occur within 200 m on both sides of the Maihar–Satna highway (SH 11).

Stop 1 Lilji Nala

This field spot is located around 2 km north of Maihar town along the SH 11, close to the St. Thomas School (C1 in Fig. 3.1). Good exposures of the Lower Bhander Sandstone occur along a river, the Lilji Nala, immediately after crossing the school. This section exposes the upper part of the Lower Bhander Sandstone constituted by conglomerate, sandstone and mudstone. The dip of the sandstone is ca. 3–4° towards NW. Bed surface primary sedimentary structures are well preserved all through the exposure. Sandstone beds exhibit excellent preservation of delicate structure like rain imprints and salt pseudomorphs (Fig. 3.15a, b). The contact between the Lower Bhander Sandstone and the Sirbu Shale occurs along the downdip direction.

Stop 2 Patna Nala

This field spot is located a few kilometre north of stop 1 along the Patna Nala (C2 in Fig. 3.1). Exposures of the Lower Bhander Sandstone and the Sirbu shale occur immediately below the bridge on Patna Nala. Large stromatolites mark the bottom of the Sirbu Shale with average column height 35 cm and head diameter 25 cm (Fig. 3.16). The stromatolitic beds marking the base of Sirbu Shale are up to 10 m thick. Stromatolitic beds contain several ooid-rich laminae. The Lower Bhander Sandstone consists of alternations between red shale and sandstone. The sandstone bed surfaces exhibit abundant desiccation cracks.

Stop 3 Beta Village

This stop is on the right side of the road after crossing the village Beta between Unchehara and Nagod (C3 in Fig. 3.1). The poorly sorted sandstone of possible tsunami origin of the Sirbu Shale is well exposed in the Nala section just beside the road. This facies is non-repetitive and is present only at this location. An unusually

Fig. 3.15 Salt pseudomorphs (**a**) and rain prints (**b**) on sandstone bed surfaces (pen length = 14 cm) within the Lower Bhander Sandstone

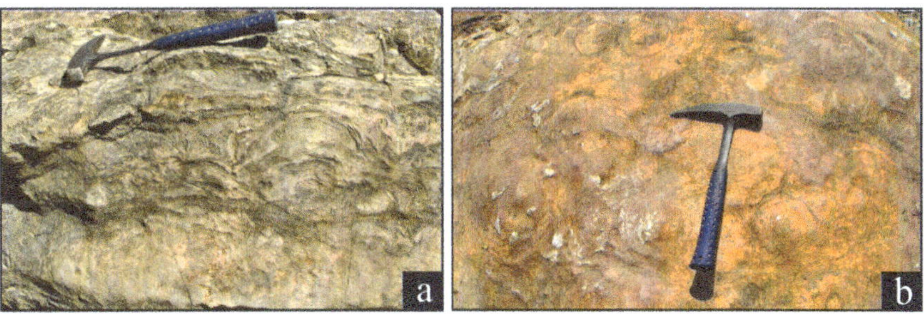

Fig. 3.16 Large-scale stromatolites (**a**) and stromatolite heads on bed surface (**b**) near the basal part of the Sirbu Shale (hammer length = 38 cm)

Fig. 3.17 Undulatory, relatively coarse sandstone bed within fine shale (**a**) hummocky cross-stratification in profile (**b**) and planar lamination in profile (**c**) within the Sirbu Shale (pen length = 14 cm)

thick (c.a 1.5 m), medium- to fine-grained sandstone is encased within the greenish-grey shale. The base of the sandstone is undulatory (Fig. 3.17a). Large-scale ripples occur on sandstone bed surfaces. The upper contact of the sandstone with the shale is also very sharp. The sandstone exhibits hummocky cross-stratification besides parallel laminae (Fig. 3.17b, c). This sandstone is laterally traceable up to half a kilometre along the river section.

3.4 Panna–Ken River

This traverse is proposed for visiting sections at the northern fringe of the Vindhyan basin around Panna (Fig. 3.18). The area is densely forested, and workable exposures

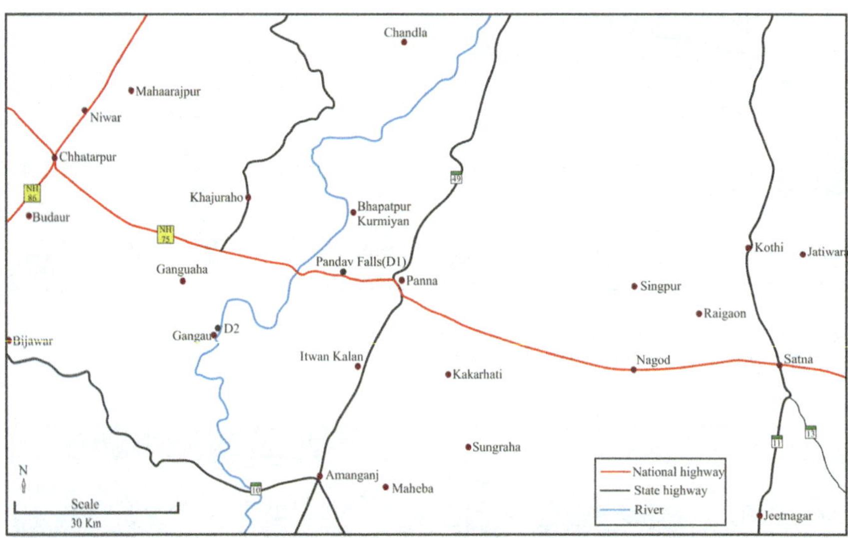

Fig. 3.18 Location map showing field spots (red dots) and localities (black dots) for the Panna–Ken River traverse

are confined along the Nala and river cuttings. The succession unconformably overlies the Bijawar Group of metasediments. Bounded between the basement below and the Kaimur Formation above the Semri Group is exposed in two good field spots, viz. the Pandwa falls and along the Ken River, near Gangau village.

Stop 1 Pandav Falls

The Pandav falls section, exposing the Semri Group, occurs north of NH 75 (D1 in Fig. 3.18). The thickness of Semri rocks is ~20 m. The characteristics of the Semri Group of rocks in this section are different from those exposed at the southern margin of the Son valley. The section exposes coarse- to medium-grained, trough cross-stratified sandstone, parallel-laminated fine-grained sandstone, mm-scale alternations between sand/siltstones and mudstone, conglomerates layers, dolostone and mudstone in a vertical section. Lag deposits occur at the base of this section. The lag conglomerate is clast- to matrix-supported, and the interstitial matrix is made up of coarse-grained sand. The clast-supported conglomerate is massive. The bottom part of the section exhibits pebbly sandstone. The medium- to coarse-grained sandstones show trough cross-stratification. Cosets of cross-strata ranging in thickness from 1.0 to 3.0 m have sharp, concave upward erosional bases and generally sharp and planar tops. Cross-set thickness within the cosets varies from 18 to 44 cm. This part is overlain by mm-scale alternations between laterally persistent, light-coloured and fine-grained sand/siltstone and dark-coloured mudstone laminae (Fig. 3.19). The thickness of the mud laminae ranges from 0.5 to 6 mm, while that of coarser clastics (sandstone/siltstone) varies between 1.5 and 9 mm. The sandstone/siltstone laminae show normal grading. This unit is followed upwards by a medium-grained,

Fig. 3.19 Alternations between laterally persistent, light-coloured and fine-grained sand/siltstone and dark-coloured mudstone laminae (hammer head for scale)

moderately sorted sandstone, internally characterized by planar tabular cross-strata. Cross-sets are solitary and are up to 1.75 m thick. A dolostone facies, buff or grey on a fresh surface and dirty yellow on the weathered outcrop, is also present in this section. The thickness of the dolostone layers varies laterally in the Pandav falls section, averaging 3.5 m. Internally dolostone layers exhibit mm- to cm-scale planar laminae. The topmost facies of this section consists of monotonous dark grey to black-coloured shale. The shale is internally planar-laminated and becomes more fissile upwards. Spherical to disc-shaped carbonate concretions occur locally within the shale. A cross-stratified Kaimur Sandstone occupies the top part of the Pandav falls section.

Stop 2 Ken River Section Around Gangau Dam

This field spot is located around Gangau village and the nearby Gangau dam (D2 in Fig. 3.18). Good exposures of lower Semri rocks occur (Gangau Formation of Mathur 1981) along the river bank and nearby areas. A diamictite body containing clasts of different sizes and compositions is exposed in this area. The diamictite facies (known as Gangwan diamictite, Ahmad 1971; Mathur 1981) represents a large channelized body that rests unconformably on the Bijawar Group. Several early investigators considered a tillite origin for the poorly sorted beds in the 'basal conglomerate' of the Semri Group (Ahmad 1958). Clasts of the diamictite may be up to 1.4 m. They include commonly vein quartz and quartzite; and less commonly jasper/chert, mica schist, dolomite, red mudstone and flattened ferruginous nodules, traceable into the underlying Bijawar Group. Lenses of massive and poorly sorted diamictites (average length and thickness 2.80 m and 40 cm, respectively) coalesce together to give rise to tabular or broadly lenticular bodies of average thickness 85 cm. Vertically juxtaposed composite lenses are separated from each other by sharp planar or curvilinear shear surfaces. Adjacent lenses differ conspicuously from each other in clast size and fabric. Faceted and striated clasts are present within this diamictite body. Broadly lenticular massive diamictite bodies locally contain strongly flattened and stretched rip-up clasts. Faceted clasts are uncommon at the upper part of the diamictite body.

The glaciogenic interpretation of the diamictite is based on its poor sorting and faceted and striated clasts. Based on these features, early workers have proposed a glacial origin for the Gangau Formation. However, later workers found it as a product of non-glacial origin (Mathur 1954, 1981; Laxmanan 1968; Ahmad 1971; Williams and Schmidt 2003).

3.5   Markandey Ghat-Kudari

It is a short traverse through the exposures of the Lower Vindhyan rocks. The traverse begins from the northern bank of the River Son near Markandey Ghat and ends in Kudari (Fig. 3.1). This traverse is parallel to the road joining Dhanwahi and Bhadanpur. Excellent outcrops of Kaimur and Rewa Formations occur along the road joining Kudari and Amarpatan, particularly along the hill range. However, to avoid repetition, we have skipped describing the field spots. Many outcrops of the basal stratigraphic units of the Semri Group in this area are drowned under water because of the construction of Bansagar Dam.

Stop 1 Markandey Ghat Section

It is an excellent section exposing the Koldaha Shale on the northern bank of the Bansagar Lake (E1 in Fig. 3.1). While the bottom part of the section is dominated by shale, the top part is sandy. The section documents a gradational transition from offshore-originated shale to coarse-grained fluvial sandstone through shale–sandstone alternation, fine-grained and medium-grained sandstones (Fig. 3.20a). While the geometry of the sandstone beds at the bottom part of the section is predominantly sheet-like, these are lenticular at the upper part. Both tabular and trough cross-stratification are abundant within the coarse sandstones at the top part. The detailed

Fig. 3.20 Sandstone–shale alternation in the Kheinjua Formation (**a**) and outcrop of Fawn Limestone showing tepee structure (**b**) (hammer length = 38 cm)

lithology of this section is provided in Fig. 2.13. The facies succession reveals an overall prograding fluvio-deltaic deposits within the Koldaha Shale. Microbial carbonates locally occur as isolated outcrops on both sides of road joining Markandey Ghat and Gorsari. These local microbial dolomite bodies within the Kheinjua Formation belong to the Fawn Limestone. These exhibit excellent tepee structures (Fig. 3.20b). Locally the dolomite bodies exhibit cherty bands within them

Stop 2 Kudari

This stop is located in the middle of the road joining Markandey and Gorsari (E2 in Fig. 3.1). The ~85 m thick Rampur Shale is exposed along a low-relief ridge. The Rampur Shale comprises three successive facies which build up a fining-upward composite succession. It lacks emergence features throughout. The basal facies is about 8 m thick and consists of buff-coloured, quartz-rich, fine-grained sandstone that is faintly cross-stratified. The cross-strata are planar, quasi-planar and rarely hummocky. Wave ripples occur on sandstone bed surfaces. Average heights and wavelengths of ripples are about 1.8 and 5.1 cm, respectively. About 60 m thick greenish-grey shale with isolated quartz-rich and fine-grained sandstones dominates the mid-level of the Rampur Shale. The shale is olive green in colour and contains about 1.1 wt% total organic carbon. Millimetre-scale, laterally persistent siltstone inter-laminae within the shale are either massive or faintly planar-laminated. The sandstone is mostly confined to small gutter casts except for local spillover lobes. These sandstone lenses exhibit planar, ripple and wavy laminae. They also show overall grading. Wave ripples are present locally on bed tops. Towards the top of the facies unit, siltstone-filled gutters are thinner and wider (Fig. 3.21a). Siltstones within gutter casts are generally massive- or planar-laminated. Wave ripples are less abundant, and convolute laminae are present locally. The distinctly fining-upward marine Rampur Shale, lacking emergence features, represents a transgressive systems tract (TST) deposit that rests on a highstand systems tract (HST) comprising the fluvio-deltaic Kheinjua Formation (Fig. 3.21b; see Sarkar et al. 2002b for details).

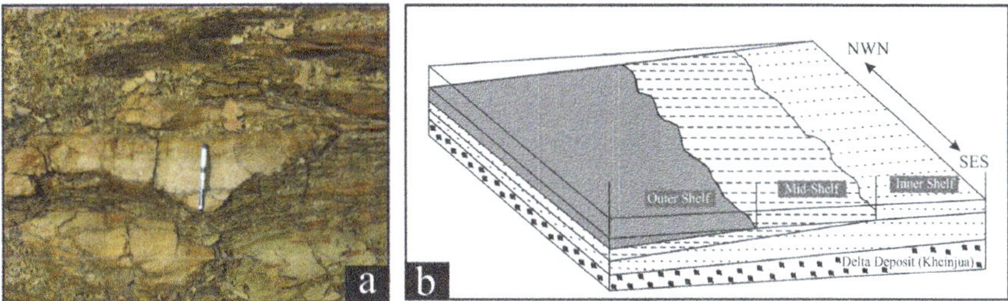

Fig. 3.21 Gutter cast in profile (**a**) and depositional model for the Rampur Shale showing transgressive facies architecture (**b**) (pen length = 14 cm)

3.6   Shikarganj–Semariya–Chorhat–Baghwar

This traverse covers the best exposures of the entire Semri Group around Chorhat and Shikarganj area.

Stop 1 Shikarganj Bridge

The basal Deoland Formation is well exposed on the eastern side of Son River bridge in Shikarnagj (E1 in Fig. 3.1) and also along low ridges southeast of Semariya. The thickness of the Deoland Formation varies from 95 m in Semariya to 120 m in Shikarganj area. The lithological assemblage of the Deoland Formation includes conglomerate, pebbly sandstone, medium-grained sandstone, with shale at the top part. Pebbly sandstones and conglomerates occur intermittently across the formation except the top 20 m. The non-pebbly, medium-grained sandstones exhibit tabular cross-stratification (Fig. 3.22). The cross-stratification orientation is multimodal, despite being dominantly northwesterly. The sandstone bed surfaces occasionally exhibit near-symmetric wave ripples. The pebbly sandstone is broadly lenticular and is internally tabular cross-stratified. Locally cross-stratifications show herring-bones. Pebbles are sprinkled in varied frequency in non-pebbly sandstones. In places, underlying and overlying laminae wrap around pebbles. Locally the pebble content exceeds 30%. The formation is overall fining upwards in this section, and it gradationally passes over to the Kajrahat Formation.

Stop 2 South of Son River Bridge, North of Badhaura

This field spot exposes the Porcellanite Formation along with a road cutting (E2 in Fig. 3.1). Since the early works of Auden (1933), this formation has been considered as dominantly pyroclastic. The presence of abundant pumice, delicate glass shards and euhedral quartz grains indicate the pyroclastic nature. The thickness of the Porcellanite Formation is approximately 160 m in this area. Tabular beds of Porcellanite with alternate black and white banding form the most conspicuous facies in this section. Dark grey beds of the Porcellanite Formation may exhibit normal grading in places. A few of the beds show eolian ripples with large wavelength: amplitude

Fig. 3.22 Medium-grained sandstone exhibiting tabular cross-stratifications within the Deoland Formation

ratio, with a concentration of coarse grains on crests. Cross-stratified porcellanite beds are occasionally present. Some of the beds are very poorly sorted, massive and fine-grained with floating clasts ranging in length up to 35 cm. Slump folds are prominent features within these beds. In some beds, massive, normally graded pyroclasts with scoured bases pass upwards to parallel lamination and climbing ripple lamination arranged in Bouma cycles consisting of divisions a, b and c. These beds are similar to subaqueous volcaniclastic turbidites. Greyish to greenish shale without any intervening siltstone and sandstone beds alternate with turbidite-like beds. Poorly sorted, brecciated beds with sharp lower and upper contacts, consisting of ash to block-sized angular clasts of altered glasses occur at the basal part of the section. Facies succession of the Porcellanite Formation reveals repeated vertical juxtaposition of subaerial and subaqueous components and gradational transition to the overlying Koldaha Shale (Banerjee 1997).

Stop 3 South of Chorhat Guest House (E3 missing in Fig. 3.1)

This section exposes the Koldaha Shale primarily with local patches of stromatolitic limestone (E3 in Fig. 3.1). The thickness of the Koldaha Shale attains up to 340 m in this area. The Koldaha Shale consists of four distinctive facies (for details, see Table 2.4). The shale facies appears dark grey and it occurs at the basal part of the Koldaha Shale. The repeated alternation between dark grey shale and sheet-like, light yellowish sandstone characterizes the heterolithic facies (Banerjee 2000). The bases of the sandstone beds are sharp and locally bear mud clasts. Internally, they exhibit hummocky cross-stratification and planar lamination. Beds are mantled by wave-cum current ripples (Fig. 3.23). Vertical thickness variations of the shale and sandstone alternations give rise to coarsening- and thickening-upward as well as fining- and thinning-upward stratal packages. Sandstone facies form westward wedging bodies at two different stratigraphic levels and remain encased within fine-grained facies (Fig. 2.9b). The sandstone bodies are several km long. Both these sandstone bodies are overall coarsening upward and bear desiccation cracks, rain prints and other emergence features at the top part. While at the lower part the sandstone beds are lenticular or wedge-shaped, at the upper part the constituent beds are internally cross-stratified. These two sandstone bodies are considered as fluvio-deltaic wedges (Sarkar et al. 1996; Bose et al. 1997, 2001; Banerjee 2000). Lenticular bodies of mixed carbonate and siliciclastic lithologies occur locally, containing crinkly laminated microbial laminae and stromatolites.

Fig. 3.23 Wave-cum current ripples on a sandstone bed surface within Rampur Shale (pen length = 14 cm)

Stop 4 Chorhat–Rampur Nekin Road

It includes a series of stops along a straight road from Chorhat to Rampur Nekin (E4 in Fig. 3.1) on the northern side of NE–SW aligned ridge that exposes the shallow marine originated Chorhat Sandstone, with superbly preserved wave/current originated features and microbial mat structures. The Chorhat Sandstone, the upper member of the Kheinjua Formation, is best exposed in this area, consisting dominantly of shallow marine sandstones. Dominantly sandy though, the member also contains thin red-coloured mud partings at places. The spectacular preservation of wave ripples exhibiting widely variable pattern is a special attraction of this section. Microbially related structures are abundant in almost every sandstone bed surfaces (see Chap. 5). Indeed, a wide variety of mat-related structures have recently been reported from this section (Sarkar et al. 2004b, 2005, 2006, 2014b, 2016; Sakar and Banerjee 2007; Bose et al. 2007; Banerjee and Jeevankumar 2005; Banerjee et al. 2010, 2014; for details see Chap. 5). Seilacher et al. (1998) reported traces of potential undermat burrowers from the Chorhat Sandstone. The vertical stacking of 35–40 cm thick beds of light-coloured, overall graded, tabular sandstone, alternating with siltstone less than 5 cm thick dominates the lower part of the Chorhat Sandstone. The bottom of the sandstone beds is riddled with gutters and prod marks, while internally the sandstone beds exhibit hummocky cross-stratification, quasi-planar strata and wave ripples. Well-sorted, fine-grained, wave-rippled sandstone with emergence features occur towards the top, which is associated with well-sorted sandstone with various eolian features. A progradational succession, built upon a storm-dominated shelf, has been inferred for the Chorhat Sandstone. The Chorhat Sandstone is overall fining-upward, and it sharply passes upwards to the Rampur Shale.

Stop 5 Rampur Guest House

Organic-rich shale belonging to the Rampur Shale Member of the Rohtas Formation crops out in a small rivulet section near Rampur Guest House (E5 in Fig. 3.1). Highly organic-rich shale (TOC up to 3.5%) alternates with thin beds of crinkly laminated limestone (Fig. 3.24). The thickness of the Rampur Shale is around 20 m in this area. Although the black shale is hard and compact, under the microscope it reveals a broad spectrum of microbial mat-related structures including wavy and crinkly carbonaceous laminae and pyritic laminae (Sur et al. 2006; Banerjee et al. 2006b; Schieber et al. 2007). The limestone bed thickness increases, while black shale reduces upwards as the Rampur Shale gradationally passes over to the Rohtas

Fig. 3.24 Alternations between black shale and limestone within Rampur Shale (hammer length = 38 cm)

Fig. 3.25 Ribbon limestone (**a**) rippled limestone (**b**), edgewise and non-edgewise limestone conglomerates (**c** and **d**) within Rohtas Limestone (Swiss knife length = 9.1 cm)

Limestone. The bedding surfaces of black shale beds show circular and tubular carbonaceous fossils, named as *Chuaria* and *Tawuia*.

Stop 6 Baghwar

Limestone mines and abandoned quarries expose the Rohtas Limestone on the southern flank of the Kaimur Range (E6 in Fig. 3.1). Although fresh exposures of the limestone mines do not reveal internal structures, abandoned quarries are amenable for detailed facies analysis. The thickness of the Rohtas Limestone is around 155 m in this section. The Rohtas Limestone consists of ribbon limestone with alternate light and dark bands of calcimicrite, thin beds of calcisiltite, wavy-laminated calcarenites, edgewise and non-edgewise limestone conglomerates (calcirudites) (Fig. 3.25). The Rohtas Limestone is overall progradational.

3.7 Obra–Dala–Ghurma

This traverse covers the exposures of the entire Semri Group in eastern Son valley (Fig. 3.26). This traverse is excellent for good outcrops of Deoland, Kajrahat, Porcellanite and Rohtas Formations.

Stop1 South of Obra–Rihand River

The entire Deoland Formation is well exposed on the west bank of River Rihand and south of Obra town (G1 in Fig. 3.26). The lower part of the Deoland Formation consists of breccia, conglomerate and pebbly sandstone. The thickness of the lower part is around 110 m. The upper part of the Deoland Formation consists of chevron cross-stratified sandstone, planar-laminated sandstone, hummocky cross-stratified sandstone, trough cross-stratified sandstone, siltstone and shale (Fig. 2.1).

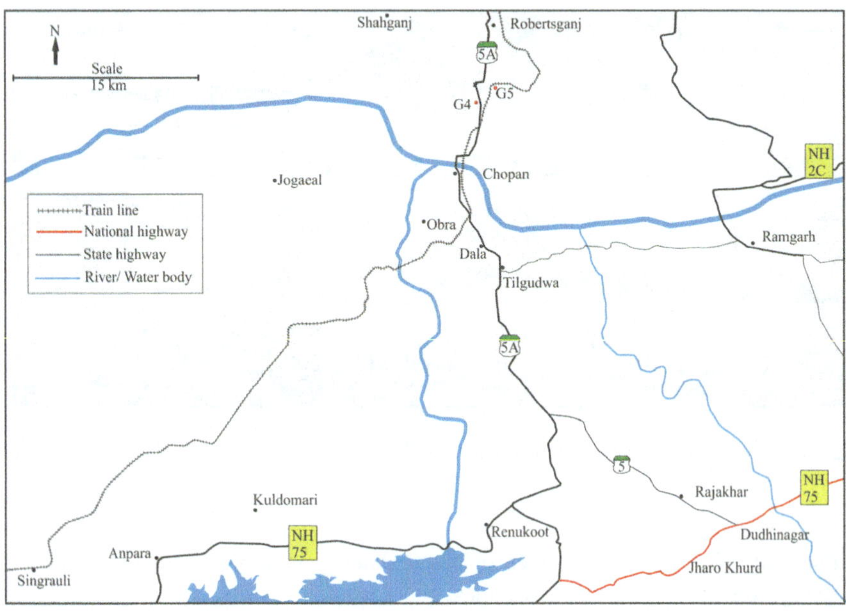

Fig. 3.26 Field spots (red dots) and localities (black dots) for the Obra–Dala–Ghurma traverse

The Deoland Formation is overall fining-upward, and it gradationally passes upwards to the Arangi Shale. In contrast to the sediments of the lower part, those of the upper part are strikingly better sorted, matrix-free, and completely devoid of pebbles. The sandstones are compositionally arkose to sub-arkose, gaining mineralogical maturity upwards. Sandstones appear greenish-grey because of the presence of ferric illite (Banerjee et al. 2008; Banerjee 2010). Poor sorting of clasts and grains in the lower part indicates terrestrial deposition. Alternations between planar-laminated and cross-stratified sandstone at the upper part suggest deposition in shoreface conditions. Chevron cross-stratified beds indicate vertically accreted shoals. The fining-upward trend at the upper part of the Deoland Formation and the upward transition into the Arangi Shale clearly indicates the rise in relative sea level at a rate exceeding the sedimentation rate.

Stop 2 East of Dala Cement Factory

The bottom part of the Kajrahat Limestone is well exposed near Dala cement factory (G2 in Fig. 3.26). This part of the Kajrahat limestone is around 12.5 m thick and is well exposed in isolated quarries. It consists of black shale, microbial-laminated dolomite and planar-laminated dolomite. Black shale at the bottom part of the Kajrahat Limestone is organic-rich with TOC content varying from 1.1 to 3.9%. Tabular bed (av. thickness 10 cm) of microbial-laminated and planar-laminated dolomite alternates with black shale. Carbonaceous discs of average diameter 2.5 mm can be observed on bedding surfaces of black shales in places. The shale is pyritiferous in places. Wavy–crinkly and carbonaceous laminae with abundant pyrite framboid are typical features of the shale under the microscope (Banerjee et al. 2006b; Sur et al. 2006; Schieber et al. 2007). The middle part of the Kajrahat Limestone (~95 thick) solely consists of microbial-laminated dolomite around Obra area. While the top part of the Kajrahat Limestone exhibits cyclic alternations between large and small stromatolites, black shale at the bottom of the Kajrahat Limestone has been considered as indigenous offshore mud deposits. The planar-laminated dolomite represents possible storm deposits in distal shelf environment. The storm interpretation is supported by the sharp base, planar and convolute laminae, and occasional graded bedding. The crinkly laminated dolomite resembles microbial laminite (Schieber et al. 2007). The abundance of small stromatolites at the upper division suggests the overall progradational characteristics of the Kajrahat Limestone.

Stop 3 Chopan Railway Cutting

The Porcellanite Formation is well exposed along the railway cutting south of Son bridge of Chopan (G3 in Fig. 3.26). The thickness of the Porcellanite Formation varies from 70 to 105 m in this area. Facies constituting the formation include alternate black and white banded, crudely stratified, cross-stratified, shale, clast-rich and graded porcellanites. While alternating dark and white banded variety of porcellanite represents periodic hot and cold ash deposition, other varieties suggest subaqueous deposition. Roy and Banerjee (2002) reported recurrent facies transitions between subaqueous and subaerial facies (see also Srivastava et al. 2003).

Stop 4 Southeast of Salkhan

The Fawn Limestone is nicely exposed in this section southeast of Salkhan village (G4 in Fig. 3.27). The maximum thickness of the Fawn Limestone in this area is up to 90 m. Stromatolites and microbial laminites are the major constituents of the Fawn Limestone (Fig. 3.26) The microbial-laminated dolomite exhibits prominent bird's eye (fenestrae) structures in the field.

Fig. 3.27 Stromatolites heads in the bedding plane section of the Salkhan Limestone (coin diameter = 2.7 cm)

Stop 5 Ghurma

Abandoned quarries of the Ghurma limestone mines expose the complete section of the Rohtas Limestone (G5 in Fig. 3.26). The approximately 85 m-thick Rohtas Limestone consists of seven facies in this area (Fig. 3.28; for details, see Banerjee et al. 2005, 2006a; Banerjee and Jeevankumar 2007). Organic-rich black shale and crinkle-laminated limestone dominates the basal part, reveals the distinctive fabric of microbial mat origin and represents outer shelf products. Crinkle-laminated limestone facies gains importance and locally turns into the nodular limestone facies through early diagenetic modification in the middle part. The grey shale facies, poor in organic content, replace black shale at the upper part. The top part of the succession consists of heterolithic facies, planar-laminated limestone facies and wavy-laminated limestone facies alternating with the grey shale facies. While the heterolithic facies records deposition across the paleochemocline, the planar-laminated and the wavy-

Fig. 3.28 Salient features of the Rohtas Limestone: Organic-rich black shale (above the pen) overlying crinkle-laminated limestone (**a**), crinkle-laminated limestone (**b**) heterolithic facies (**c**), planar-laminated limestone facies (**d**), alternation between limestone and grey shale (**e**) (pen length = 14 cm)

laminated limestone facies document deposition in a shallow agitated shelf. Confined between a condensed zone below and an unconformity above, the Rohtas Limestone is overall coarsening-upward.

Coordinates of Locations Mentioned in This Chapter

Dhanwahi	23°59′29.45″N; 80°47′36.33″E
Kuteswar	23°59′1.93″N; 80°48′45.82″E
Guguwar	24°4′23.59″N; 80°48′11.19″E
Bhadanpur	24°9′35.47″N; 80°49′2.51″E
Kaimur Ghat	24°10′55.06″N; 80°48′45.31″E
Sarlanagar	24°12′4.47″N; 80°48′6.29″E
Hathi Nala	24°12′22.13″N; 80°48′22.40″E
Thomas River	24°13′34.70″N; 80°47′36.53″E
Girgita	24°15′28.89″N; 80°48′18.29″E
Emilia	24°15′28.89″N; 80°48′18.29″E
Dolni	24°18′10.72″N; 80°48′11.49″E
Sarda Temple	24°15′40.12″N; 80°43′22.70″E
Rampur	24°17′28.67″N; 80°40′57.61″E
Maihar	24°16′9.89″N; 80°45′23.94″E
Lilji Nala	24°16′53.05″N; 80°45′39.28″E
Patna Nala	24°18′35.45″N; 80°46′9.42″E
Beta	24°25′41.22″N; 80°46′51.86″E
Unchehara	24°23′42.86″N; 80°47′46.72″E
Panna	24°43′5.04″N; 80°10′54.72″E
Pandawa falls	24°44′16.20″N; 80° 0′38.46″E
Gangau dam	24°35′18.43″N; 79°49′57.78″E
Shikarganj	24°17′13.33″N; 81°27′48.86″E
Badhaura	24°23′48.75″N; 81°44′35.31″E
Chorhat Guest House	24°25′3.40″N; 81°40′25.89″E
Rampur-Nekin	24°20′32.32″N; 81°28′33.45″E
Rampur Guest House	24°19′59.28″N; 81°26′43.87″E
Baghwar	24°19′52.17″N; 81°24′31.14″E
Markandey Ghat	24°5′39.29″N; 81°0′28.35″E
Kudari	24°7′52.86″N; 81°0′43.07″E
Obra	24°28′28.29″N; 82°59′15.99″E
Dalla	24°27′12.24″N; 83°2′29.56″E
Chopan	24°31′23.78″N; 83°1′12.12″E

(continued)

(continued)

| Salkhan | 24°33′34.89″N; 83°2′29.44″E |
| Ghurma | 24°37′16.62″N; 83°5′42.80″E |

References

Auden JB (1933) Vindhyan sedimentation in the Son Valley, Mirzapur district. Mem Geol Surv India 62:140–250

Ahmad F (1958) Paleogeography of central India in the Vindhyan period. Rec Geol Surv India 87:513–548

Ahmad F (1971) Geology of the Vindhyan system in the eastern part of the Son Valley in Mirzapur district, U.P. Rec Geol Surv India 96:1–41

Banerjee S (1997) Facets of Mesoproterozoic Semri sedimentation, Son valley, MP. Unpubl PhD Thesis, Jadavpur University, Kolkata, India

Banerjee S (2000) Climatic versus tectonic control on storm cyclicity in Mesoproterozoic Koldaha Shale, central India. Gond Res 3:521–528

Banerjee S (2010) Distinction between marine and continental facies in Precambrian sedimentary succession: Palaeoproterozoic Deoland Formation, Vindhyan Supergroup, central India. Gond Geo Mag 25:239–250

Banerjee S, Jeevankumar S (2005) Microbially originated wrinkle structures on sandstone and their stratigraphic context: Palaeoproterozoic Koldaha Shale, central India. Sed Geol 176:211–224

Banerjee S, Jeevankumar S (2007) Facies and depositional sequence of the Mesoproterozoic Rohtas Limestone: Eastern Son valley, India. J Asian Earth Sci 30:82–92

Banerjee S, Sarkar S, Bhattacharyya SK (2005) Facies, dissolution seams and stable isotope characteristics of the Rohtas Limestone (Vindhyan Supergroup) in the Son valley area, central India. J Earth Sys Sci 114:87–96

Banerjee S, Jeevankumar S, Sanyal P, Bhattacharyya SK (2006a) Stable isotope ratios and nodular limestone of the Proterozoic Rohtas Limestone: Vindhyan Basin, India. Carb Evap 21:133–143

Banerjee S, Dutta S, Paikaray S, Mann U (2006b) Stratigraphy, sedimentology and bulk organic geochemistry of black shales from the Proterozoic Vindhyan Supergroup (central India). J Earth Sys Sci 115:37–48

Banerjee S, Bhattacharya SK, Sarkar S (2007) Carbon and oxygen isotopic variations in peritidal stromatolite cycles, Paleoproterozoic Kajrahat Limestone, Vindhyan Basin of central India. J Asian Earth Sci 29:823–831

Banerjee S, Jeevankumar S, Eriksson PG (2008) Mg–rich ferric illite in marine transgressive and highstand systems tracts: examples from the Paleoproterozoic Semri Group, central India. Precam Res 162:212–226

Banerjee S, Sarkar S, Eriksson PG, Samanta P (2010) Microbially related structures in siliciclastic sediment resembling Ediacaran fossils: examples from India, ancient and modern. In: Seckbach J, Oren A (eds) Microbial mats: modern and ancient microorganisms in stratified system. Springer-Verlag, Berlin, pp 111–129

Banerjee S, Sarkar S, Eriksson PG (2014) Palaeoenvironmental and biostratigraphic implications of microbial mat–related structures: examples from modern Gulf of Cambay and Precambrian Vindhyan basin. J Paleogeography 3:127–144

Banks NL (1973) Falling stage features of a Precambrian braided stream: criteria for subaerial exposure. Sed Geol 10:147–154

Bose PK, Chakraborty PP (1994) Marine to fluvial transition: Proterozoic Upper Rewa sandstone, Maihar, India. Sed Geol 89:285–302

Bose PK, Banerjee S, Sarkar S (1997) Slope–controlled seismic deformation and tectonic framework of deposition of Koldaha Shale, India. Tectonophys 269:151–169

Bose PK, Chakraborty S, Sarkar S (1999) Recognition of ancient aeolian longitudinal dunes: a case study from the Upper Bhander Sandstone, Son Valley, India. J Sed Res 69:86–95

Bose PK, Sarkar S, Chakraborty S, Banerjee S (2001) Overview of the Meso to Neoproterozoic evolution of the Vindhyan basin, central India. Sed Geol 141:395–419

Bose PK, Sarkar S, Banerjee S, Chakraborty S (2007) Mat–related features from the Vindhyan Supergroup in central India. In: Schieber J, Bose PK, Eriksson PG, Banerjee S, Sarkar S, Catuneanu O, Altermann W (eds) An atlas of microbial mat features preserved within the clastic rock record. Elsevier, pp 181–188

Chakraborty PP, Sarkar S, Bose PK (1998) A viewpoint on intracratonic chenier evolution: Clue from a reappraisal of the Proterozoic Ganurgarh Shale, Central India. In: Paliwal BS (ed) The Indian Precambrian. Scientific Publishers, Jodhpur, pp 61–72

Cotter E (1978) The evolution of fluvial style, with special reference to the central Appalachian Paleozoic. In: Miall AD (ed) Fluvial sedimentology. Can Soc Petrol Geol Mem 5:361–383

de Raaf JFM, Boersma JR, Van Gelder A (1977) Wave generated structures and sequences from a shallow marine succession. Lower Carboniferous County Cork. Ireland. Sedimentol 24:451–483

Fuller AO (1985) A contribution to the conceptual modelling of pre-Devonian fluvial systems. Trans Geol Soc South Africa 88:189–194

Kreisa RD, Moila RJ (1986) Sigmoidal tidal bundles and other tide-generated sedimentary structures of the Curtis Formation, Utah. Geol Soc America Bull 97:381–387

Laxmanan S (1968) On the nature of basal conglomerate of the Semri Series on the Son valley. Proc Nat Inst Sci India A34:50–55

Long DGF (1978) Proterozoic stream deposits: some problems of recognition and interpretation of ancient sandy fluvial systems. In: Miall AD (ed) Fluvial Sedimentology. Can Soc Petrol Geol Mem 5:313–341

Mathur SM (1954) Late Precambrian glaciation in Central India-rejoinder. Curr Sci 23:7–8

Mathur SM (1981) The middle Proterozoic Gangau tillite, Bijawar Group, Central India. In: Hambrey MJ, Harland WB (eds) Earth's pre-pleistocene glacial records. Cambridge University Press, Cambridge, pp 424–427

Roy S, Banerjee S (2002) Facies and petrography of the Porcellanite Formation around Chopan, Uttar Pradesh. J Indian Ass Sedimentol 20:195–205

Sarkar S, Banerjee S (2007) Some unusual and/or problematic features. In: Schieber J, Bose PK, Eriksson PG, Banerjee S, Sarkar S, Catuneanu O, Altermann W (eds) An atlas of microbial mat features preserved within the clastic rock record. Elsevier, Amsterdam, pp 145–147

Sarkar S, Banerjee S, Bose PK (1996) Trace fossils in the Mesoproterozoic Koldaha Shale, Central India, and their implications. Neues Jahrbuch fur Geologie und Palaontologie-Monatshefte 7:425–438

Sarkar S, Chakraborty S, Banerjee S, Bose, PK (2002a) Facies sequence and cryptic imprint of sag tectonics in late Proterozoic Sirbu Shale, central India. In: Altermann W, Corcoran P (eds) Precambrian sedimentary environments: a modern approach to ancient depositional systems. Spec Publ IAS 33, Blackwell Science, pp 369–382

Sarkar S, Banerjee S, Chakraborty C, Bose PK (2002b) Shelf storm flow dynamics: insight from the Mesoproterozoic Rampur Shale, central India. Sed Geol 147: 89–104

Sarkar S, Eriksson PG, Chakraborty S (2004a) Epeiric sea formation on Neoproterozoic supercontinent break-up: a distinctive signature in coastal storm bed amalgamation. Gond Res 7:313–322

Sarkar S, Banerjee S, Eriksson PG (2004b) Microbial mat features in sandstone illustrated. In: Eriksson PG, Altermann W, Nelson W, Mueller DR, Catuneanu O (eds) The Precambrian Earth: tempos and events. Elsevier, Amsterdam, pp 673–675

Sarkar S, Banerjee S, Eriksson PG, Catuneanu O (2005) Microbial mat control on siliciclastic Precambrian sequence stratigraphic architecture: examples from India. Sed Geol 176:195–209

Sarkar S, Banerjee S, Samanta P, Jeevankumar S (2006) Microbial mat–induced sedimentary structures and their implications: examples from Chorhat Sandstone, MP India. J Earth Sys Sci 115:49–60

Sarkar S, Bose PK, Eriksson PG (2011) Neoproterozoic tsunamite: upper Bhander Sandstone, central India. Sed Geol 238:181–190

Sarkar S, Choudhuri A, Banerjee S, van Loon AJ (Tom), Bose PK (2014a) Seismic and non-seismic soft-sediment deformation structures in the Proterozoic Bhander Limestone, central India. Geologos 20:79–93

Sarkar S, Banerjee S, Samanta P, Chakraborty N, Chakraborty PP, Mukhopadhyay S, Singh A K (2014b) Microbial mat records in siliciclastic rocks: examples from Four Indian Proterozoic basins and their modern equivalents in Gulf of Cambay. J Asian Earth Sci 91:362–377

Sarkar S, Choudhuri A, Mandal S, Eriksson PG (2016) Microbial mat related structures shared by both siliciclastic and carbonate formations. J Palaeogeography 5:278–291

Schieber J, Bose PK, Eriksson PG, Banerjee S, Sarkar S, Catuneanu O, Altermann W (2007) An Atlas of microbial mat features preserved within the siliciclastic rock record. Elsevier Science, Amsterdam

Schumm SA (1968) Speculations concerning paleohydraulic controls of terrestrial sedimentation. Geol Soc America Bull 79:1573–1588

Seilacher A (1982) Distinctive features of sandy tempestites. In: Einsele G, Seilacher A (eds) Cyclic and event stratification. Springer, Berlin, Heidelberg, pp 333–349

Seilacher A, Bose PK, Pflüger F (1998) Triploblastic animals more than 1 billion years ago: trace fossil evidence from India. Science 282:80–83

Srivastava RN, Srivastava AK, Singh KN, Redcliffe RP (2003) Sedimentation and depositional environment of the Chopan Porcellanite Formation, Semri Group, Vindhyan Supergroup in parts of Sonbhadra district, UP. J Palaeontol Soc India 48:167–179

Sur S, Schieber S, Banerjee S (2006) Petrographic observations suggestive of microbial mats from Rampur Shale and Bijaigarh Shale, Vindhyan basin, India. J Earth Sys Sci 115:61–66

Williams GE, Schmidt PW (2003) Possible fossil impression in sandstone from the late Paleo-proterozoic—early Mesoproterozoic Semri Group (lower Vindhyan Supergroup), central India. Alcheringa 27:75–76

Chapter 4
Selected Field Sections

4.1 Introduction

This chapter deals with several well-exposed sections, each of which may take a whole day to investigate properly. Each section covers a unique theme, including detailed facies analysis of carbonate and siliciclastic successions and description of event deposits like seismite and tsunamiite. One of the sections provides unique variations in the morphology of stromatolites, while a three-dimensional outcrop enables to identify different architectural elements within a fluvial succession. Coordinates of all these sections are provided at the end of this chapter.

4.2 Facies Analysis of the Rohtas Limestone in Bhandanpur

The Rohtas Limestone, belonging to Rohtas Formation, is the topmost member of Semri/Lower Vindhyan Group. It is one of the significant carbonate deposits within Vindhyan Supergroup. U/Pb SHRIMP technique and Pb–Pb dating technique have confirmed that it is ~1599 ± 48 Ma old (Sarangi et al. 2004; Fig. 1.3). The Rohtas Limestone is well-exposed south and east of Bhadanpur village. Several opencast mines along the Bhadanpur–Kaimur road exposes the Rohtas Limestone. Prior permission is necessary to visit the large opencast mines run by Maihar Cement. The quarry sections of different opencast mines provide an excellent opportunity for a detailed facies analysis of the Rohtas Limestone (Fig. 4.1). A detailed description of the constituent facies of the Rohtas Limestone in this area is provided as follows.

© Springer Nature Singapore Pte Ltd. 2020
S. Sarkar and S. Banerjee, *A Synthesis of Depositional Sequence
of the Proterozoic Vindhyan Supergroup in Son Valley*, Springer Geology,
https://doi.org/10.1007/978-981-32-9551-3_4

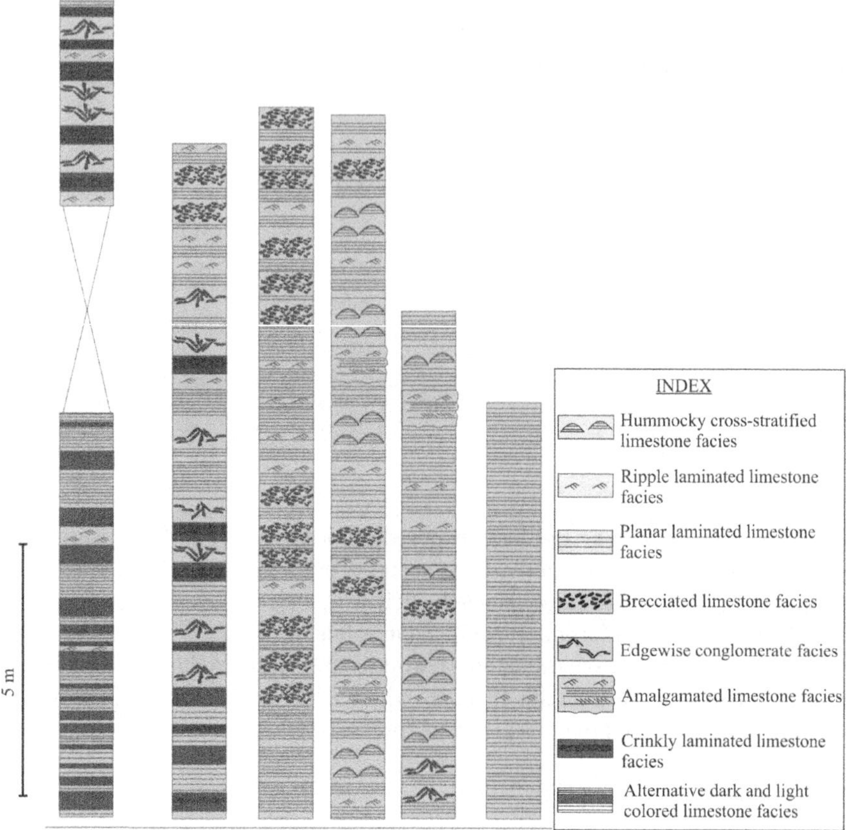

Fig. 4.1 Lithologs showing constituent facies within the Rohtas Limestone in the Bhadanpur area. Note the difference in facies distribution between the lower and upper parts of the limestone. Also, note the confinement of the alternate dark- and light-coloured limestone and crinkly laminated facies in the lower part of the succession, while the hummocky cross-stratified and planar-laminated facies are frequently present in the upper part

4.2.1 Alternate Dark and Light Coloured Limestone Facies

This facies is restricted to the basal part of the section in this area. It is very well exposed in the Maihar Cement Mines. It is characterized by laterally persistent, alternating dark and light laminations (Figs. 4.1 and 4.2a). The average thickness of the individual dark and light layers is ~2.5 cm and ~4 cm, respectively. Both the lower and upper contacts of these dark and light laminae are sharp. The dark layers are massive and micritic, whereas light laminae are microsparry in nature and show planar lamination. Rarely small-scale ripple cross-laminae are preserved within the light laminae. In a few places, small little mound-like structures are present on the

Fig. 4.2 Sedimentary features present within the Rohtas Limestone around Bhadanpur area: alternating dark- and light-coloured laminae (**a**), mound-like structures with radial cracks (**b**), amalgamated limestone with sharp and irregular bases (**c**) and planar-laminated limestone (**d**) (Swiss knife length = 9.1 cm, hammer length = 38 cm)

bedding surfaces of the dark layers (Fig. 4.2b). Radiating cracks occur at the top of the mound-like structures.

Laterally persistent and undisturbed alternate dark and light layers indicate that deposition took place in a calm and quiet environment (Sarkar et al. 2014). The absence of current structures also supports this contention. Small-scale ripple laminae within light layers develop only during rare high energy events within a low-energy environment. The presence of pyrite grains along the dark layers indicates that the depositional environment was anoxic and restricted. The mound-like structures present on the bedding surfaces of the dark layers are comparable with the gas dome structures (Bouougri and Porada 2007; Bottjer and Hagadorn 2007), and therefore those bulbous structures are possibly generated due to gas escape. Radiating cracks at the peak of the mound-like structures, overall massiveness and upward buckling nature of the dark layers also corroborate this contention (Frey et al. 2009). Possibly, these dark layers are comparable to the microbial mat layers (Schieber 2004) and gas must have been generated due to decay of these mat layers (Bottjer and Hagadorn 2007).

4.2.2   Crinkly Laminated Limestone Facies

This facies is associated with the previous facies and also occur at the basal part of the sections. The facies is laterally persistent and is characterized by mm-scale crinkled laminae. Ripple-laminated facies overlies the crinkled laminae occasionally.

Crinkly lamination indicates that these might be microbial laminites (Banerjee et al. 2007). The facies must have been deposited in a calm and quiet environment, but definitely within the photic zone.

4.2.3   Amalgamated Limestone Facies

This facies consists of amalgamated calcarenite beds having tabular body geometry (Fig. 4.2c). The maximum thickness of the bed is up to 35 cm. These amalgamated units are confined to the upper part of the succession only. The base of the beds is either sharp or highly irregular. Gutters and flutes are present on the sole of the beds. Internally, the bed is constituted by massive part at the base followed by parallel lamination. The top part of the beds shows ripple cross-lamination. At places, crude cross-stratification has also been developed above the planar-laminated unit which again gives way to wavy lamination and ripples at the top.

The occurrence of amalgamated calcarenite beds juxtaposed one above another in a single stratigraphic interval suggests rapid recurrence of event flow, possibly storm surges, in a high flow regime (Sarkar et al. 2004). Sharp, irregular base with gutters and tool marks at the sole of the bed also supports the high energy events responsible for the deposition of the beds (Sarkar et al. 2012). Planar lamination giving way upwards to ripple lamination suggests gradual waning of the high energy events (Bose and Sarkar 1991; Bose et al. 2012).

4.2.4   Planar-Laminated Limestone Facies

This facies is dominantly present at the upper part of the succession, and it is made up of thin-bedded lime mudstone (Fig. 4.2d). Within this lime mudstone unit, some thin laterally persistent calcarenite beds are present. The lower contact of the calcarenite unit is sharper than the upper one. Occasionally small-scale flute casts and prod marks are present at the base of these calcarenite units. Internally, both lime mudstone and calcarenites are mostly planar-laminated.

The lime mudstone-dominated facies represents deposition within the relatively deeper part of the shelf, definitely below the fair weather wave base. On the other hand, the laterally persistent calcarenite layers suggest deposition from storm events. The presence of sole features, e.g. flute casts, prod marks support this contention.

4.2.5 Hummocky Cross-Stratified Limestone Facies

This facies is also made up of calcarenite. The facies has very sharp bases and gradational tops. The sole of the beds is often sculpted with flute casts. Internally, it shows hummocky cross-stratification (Fig. 4.3a). It grades upwards into wave ripple-laminated limestone facies. The average wavelength and amplitude of the hummocks are 45 cm and 18 cm, respectively. The facies is present mostly towards the upper part of the succession and associated with planar-laminated limestone facies.

The presence of hummocky cross-stratification indicates deposition under the influence of oscillatory flow, possibly during storm surges. Flute casts at the sole of the beds also support this contention. The gradual transition to wave ripple-laminated limestone facies suggests gradual waning of the storm surges.

4.2.6 Brecciated Limestone Facies

This facies is composed of chaotically arranged clasts, mostly derived from the dark layers of alternate dark and light banded limestone facies (Fig. 4.3b–f). The breccias are mostly matrix-supported. The breccia bodies have wedge-shaped geometry and are variable in thickness (maximum ~45 cm). The facies has a curved base and an irregular top. The clasts are angular and haphazardly oriented. Few brecciated bodies have steeply inclined clasts, and the clasts are protruded, making the bed surface irregular. Few deformed clasts are also preserved. It is abundantly present towards the upper part of the section.

The facies is likely to be of debris flow origin (Enos 1977). Chaotically arranged, steeply inclined clasts and clasts protrusion on the bed surface indicate that the matrix strength of the flow was high (Bose and Sarkar 1991). Large, angular and haphazardly oriented clasts reclining at a high angle to the bedding indicate that the debris tumbled down along steep slopes.

Fig. 4.3 Sedimentary features within the Rohtas Limestone: hummocky cross-stratification (**a**), carbonate breccias (**b**–**f**), ripple-laminated limestone (**g**)

4.2.7   Ripple-Laminated Limestone Facies

This facies is made up of calcarenite and internally characterized by wave ripple laminae (Fig. 4.3g). The facies has tabular body geometry. This ripple-laminated limestone facies is present throughout the succession and is commonly associated with planar-laminated limestone facies. Wave ripple forms are often preserved in the bedding surfaces (wavelength ca 5 cm and amplitude ca 4 mm). The ripples are generally symmetric with a current component and often show bifurcation of ripple crests on the bedding plane.

Preserved wave ripple forms on the bedding surface confirm the presence of wave activity of the depositional regime.

4.3   Facies Analysis of Lower Bhander Sandstone in Lilji River

This section is close to Maihar town on the Maihar–Satna bus route. This section is famous for the preservation of various primary sedimentary structures including salt pseudomorphs (Fig. 4.4a, b). The Lower Bhander Sandstone Member of the Bhander Formation is well exposed in this section (Fig. 4.4c). In general, the Lower Bhander Sandstone is mud-rich; the exposure is thus limited and patchy. The mud-rich sandstone of ca 20 m thick, in this Lilji River, is usually characterized by red-coloured mudstone and tan-coloured fine-grained sandstone. At least four facies can easily be identified in this section within the Lower Bhander Sandstone, viz. mudstone, laminated sandstone, massive sandstone facies and mud-pebble conglomerate.

The mudstone facies is characterized by finely planar-laminated red-coloured mudstone and tabular bed geometry. Individual beds range in thickness from 0.2 to 65 cm, and the average is around 10 cm. The desiccation cracks are abundantly present in successive mud layers within this facies.

The laminated sandstone facies is characterized by tabular, fine- to medium-grained and tan-coloured. The sandstone beds are laterally persistent and amalgamation is common. Base of the beds is sharper than their tops. The thickness of individual beds ranges from 0.5 to 9 cm. This facies is internally characterized by a variety of laminated structures, the most common of which is planar-laminated cross-lamina. The top of sand beds invariably exhibits wave ripples. Salt pseudomorphs occur profusely at the base of the sandstone beds (Fig. 4.4a). Different types of sole features such as tool marks, prod marks, brush marks and groove marks casts are well preserved at the sole of the sand beds (Fig. 4.4d). A large variety of ripple forms such as flat-crested, superimposed and ladder-back ripples are preserved on the bed surfaces (Fig. 4.5a–d).

The massive sandstone facies is greyish white, medium-grained, lenticular-bedded, and is devoid of any internal structure. This facies is the least repetitive

Fig. 4.4 Structures of Lower Bhander Sandstone in Lilji Nala section: salt pseudomorphs under a sandstone bed (**a**), desiccation cracks on mudstone facies (**c**) and different types of sole features at the bottom of a sandstone bed (**d**) (hammer length = 38 cm, coin diameter = 2.5 mm). Lithological succession of the Lower Bhander Sandstone in Lilji Nala is provided in Fig. 4.4b

among all the facies present in this section. The thickness of the individual bodies ranges from 2 to 6 cm.

The conglomerate facies is also internally massive and have lenticular geometry. The clasts in this facies are flat red mud clasts (Fig. 4.5e). The mud clasts commonly elliptical or rod-shaped, maximum 4.5 cm in length and do not show any preferred imbrication or any evidence of sorting by size. The lower contact of the conglomerate beds is sharper than their upper contacts. The rock is generally matrix-supported; fine sand fills the interstitial spaces between the clasts. The thickness of individual beds varies from 3.5 to 25 cm. At places, it grades both laterally and vertically upwards to the massive sandstone. Wrinkle structures are locally abundant on top of sandstone bed surfaces (Fig. 4.5f). Both these facies exhibit abundant slide plains on bed surfaces.

The Lower Bhander Sandstone, constituted by four major facies, are products of different episodes. The red mudstone is authigenic, while the laminated sandstone is allogenic, possibly been deposited from the storm-induced flow, and the conglomerates and massive sandstones comprise an autokinetic population (Bose et al.

Fig. 4.5   Primary sedimentary structures on bed surfaces of Lower Bhander Sandstone in Lilji River: multidirectional ripple marks (**a**), ladder-back ripples (**b**), rain imprints (**c**), interference ripples (**d**), lenticular bodies of mud clast-bearing conglomerate (**e**), wrinkle structures (**f**) (Swiss knife length = 9.1 cm, pen length = 14 cm)

2001). A micro-tidal coastal zone where the rate of evaporation was extremely high was the likely depositional site of mudstone facies. Authigenically deposited red mudstone with numerous desiccation cracks supports this contention. The allogenic-laminated sandstones with sharp lower and gradational upper contacts, abundant sole features, an amalgamation of sandstone beds of sheet-like geometry, indicate definite storm intervention within the coastal playa. The autokinetic conglomerates and the associated massive sandstones are inferred as a product of debris flow which had been generated because of sliding. The slide planes associated with these facies are penecontemporaneous in origin.

4.4 Variability in Stromatolite Morphology Within Bhander Limestone in Amiliya

Thomas River section near Amiliya village is an excellent section to examine the variability in stromatolite morphology within the Bhander Limestone. It is ca 4.5 km east of the Maihar town. The road from the Maihar can directly lead to the riverbed. Besides stromatolites, planar-laminated limestone facies and breccia/conglomeratic facies are also present at the base of the section (Fig. 4.6a–b). Molar tooth structures are present within the planar-laminated limestone (Fig. 4.6c). The laminations are laterally persistent and Chert nodules occur in places.

Fig. 4.6 Characteristic features of the Bhander Limestone in Girgita and Amiliya mines: planar lamination (**a**), breccia (**b**), molar tooth structure (**c**) large stromatolite (**d**), small stromatolite (**e**), stromatolitic mound above the planar-laminated mudstone facies, (**f**) stromatolite showing branching (**g**) and crinkle lamination (**h**) (pen length = 14 cm, hammer length = 36 cm)

Broadly two types of stromatolites occur in this section, viz. large cabbage-headed biohermal type and columnar variety (Fig. 4.6d–g). The large cabbage-headed stromatolites lie immediately above the planar-laminated limestone facies (Fig. 4.6d). The diameter of these stromatolites varies from 90 to 440 cm. The individual lamina of the stromatolites is wavy in nature. Occasionally, wave ripples are also present on the top surface of the stromatolites. The cabbage-headed stromatolites are associated with desiccation cracks.

The columnar stromatolite overlies the large variety in the same section. The column height varies from 10 to 32 cm. These columnar forms are of two types: simple and composite. Branching is a common feature of these stromatolites (Fig. 4.6g). Some of the columnar forms are inclined (Fig. 4.7a). Above these columnar forms, small bulbous headed stromatolites are present which also show branching (Fig. 4.7b). The intercolumnar area of the columnar forms is filled with stromatolite debris; even some stromatolite chunks are present. Sub-vertical columnar form without intercolumnar area is also present in this section (Fig. 4.7c).

The presence of the stromatolites indicates abundant microbial mat growth within the photic zone (Kalkowsky 1908; Awramik 1971; Awramik and Margulis 1974; Riding 2000). The large cabbage-headed forms of stromatolites formed at relatively deeper water than the columnar and stratiform varieties (Beukes and Lowe 1989). The presence of broken stromatolite debris in the intercolumnar area indicates possible storm intervention. The presence of wave ripple supports wave agitation. The cracks present within these assemblages were studied rigorously and have been identified as syneresis in origin (Jüngst 1934; Plummer and Gostin 1981). The inclination of the stromatolite columns indicates the presence of strong currents in the depositional area (Sarkar and Bose 1992; Tosti and Riding 2017). The frequent branching of the columnar stromatolites corroborates the strong current or wave influence (Hofmann 1973). The transition from large domal forms to columnar forms and finally to small bulbous headed forms may indicate the decreasing water depth within the depositional basin.

4.5 Soft-Sediment Deformation Structures (SSDS) Within Bhander Limestone in Dolni

The Bhander Limestone in the Thomas River section near Dolni village is well exposed and suitable for the study of constituting facies/subfacies with excellent preservation of soft-sediment deformation structures (Chakraborty 2011; Sarkar et al. 2014). Three stacked seismites horizons are present in this section (Sarkar et al. 2014). The most common types of SSDS in the three layers are convolutes, widely varying in shape (Fig. 4.8a). Single and multiple types are present, with one or more lobes, symmetrical or asymmetrical with an average height of 13 cm and an average width of 28 cm. Further, there are contorted laminae and micro-faults (Fig. 4.8b). This

Fig. 4.7 Field photographs showing varied kinds of stromatolites within the Bhander Limestone in mines of Amiliya and Girgita: inclined stromatolite (**a**), bulbous headed stromatolite showing branching of columns (**b**), vertical stromatolite without intercolumnar area (pen length = 14 cm, red scale = 15 cm)

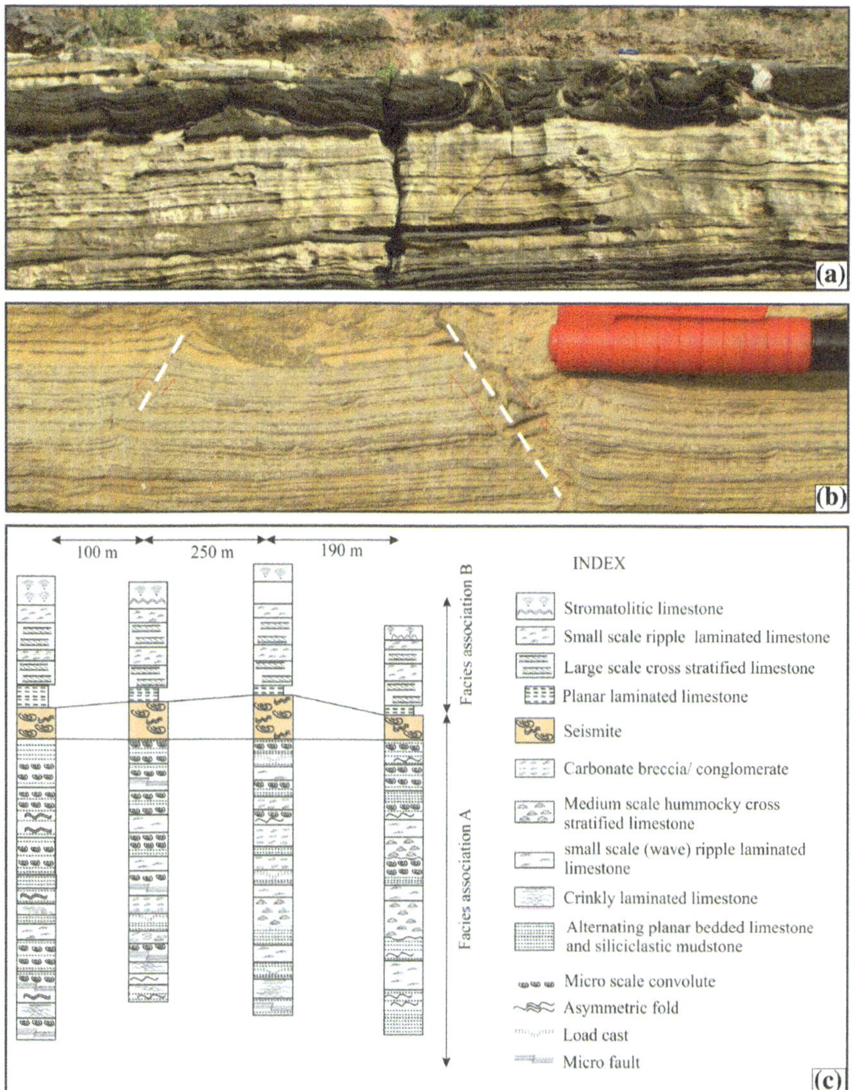

Fig. 4.8 Synsedimentary deformation structures present within Bhander Limestone: convolute lamination (**a**), micro-faults marked by white dashed lines (**b**), and lithological log showing the lateral continuity of the seismite bed (**c**) (Swiss knife length in a − 9.1 cm, pen cap length ▬ 4.5 cm)

section exhibits broadly two facies associations, below and above the seismite layers (Fig. 4.8c). The associations below and above display contrasting characteristics.

The facies association below the seismite zone consists of intraclastic and muddy carbonate made up of six facies: alternation of planar-bedded limestone and siliciclastic mudstone, small-scale (wave) ripple-laminated limestone, micro-scale convoluted limestone, medium-scale hummocky cross-stratified limestone, crinkly laminated limestone, and carbonate breccia/conglomerate (Fig. 4.9a–j). Only the first among these facies contains a significant amount of siliciclastic mud. It also shows polygonal desiccation cracks on bedding planes consisting of V-shaped cracks in vertical section. The facies association above the seismite zone comprises four distinctive facies: planar-laminated limestone, large-scale cross-stratified limestone, small-scale ripple-laminated limestone and stromatolitic limestone.

Fig. 4.9 Sedimentary features within bhander limestone: alternating planar-laminated limestone and siliciclastic mudstone (**a**), wave ripple on a bed surface (**b**), small-scale hummocky cross-stratification (**c**), crinkly laminated limestone (**d**), carbonate conglomerate/breccia (**e–f**), desiccation cracks (**g**), cross-stratification (**h**), large-scale stromatolites (**i**) and small-scale stromatolite (**j**)

4.6 Fluvial Facies Analysis Within Upper Rewa Sandstone in Chachai

This section is situated about 30 km north of the Rewa town. Positioned 5 km southwest of the Siramur village along State Highway 9, the Chachai Falls section provides unique and picturesque preservation of the Upper Rewa Sandstone Member of the Rewa Formation. A ca 120-m-deep gorge cut through the Upper Rewa Sandstone by the Rewa River, near its confluence with the Tons River, offers unique opportunity to study the stratigraphic interval in great details, including its constituting facies and architectural elements and their distribution in time and space. The now dried-up gorge, due to the construction of a barrage just upstream to the Chachai Falls, appears to be tailor-made for the studies of fluvial 'bounding surfaces', especially.

Positioned between two marine shale horizons, the Upper Rewa Sandstone Member, being fluvial in origin for the most part of it (Bose and Chakraborty 1994), represents a perfect example of progradation followed by a sharp retrogradation (Bose et al. 2001; Chakraborty 2004). The multistoried gorge section documents all the constituting facies of the Upper Rewa Sandstone. The rocks exposed here are chiefly pebbly to medium sand-sized, poorly sorted sandstone with a minor amount of fine sand to silty fraction. Petrographically the rocks are quartz arenite, with poorly sorted angular clasts of coarse-to-medium sand size. Although multistoried, the whole succession shows a more or less consistent paleocurrent direction from the bottom to the top, with a northwest-ward disposition (mean towards ~335°). State-of-the-art facies analysis identified eight genetic facies in this area.

4.6.1 Thoroughly Trough Cross-Stratified Sandstone

This facies consists of thoroughly trough cross-stratified pebbly sandstone bodies with lensoid geometry in the transverse section having concave-upward erosional bases and flat tops (Fig. 4.10a). The trough set thickness decrease upwards with no significant grain size reduction. A massive pebbly sandstone part occasionally underlies the thoroughly trough cross-stratified unit (Fig. 4.10b). Oversized clasts, ranging in size from 2 to 3 mm, are found to be dispersed along the bases of such lensoid bodies as well as along the bases of the constituting trough sets. Both the frequency and size of oversized clasts decrease up the section. Overturned cross-stratifications are quite common along with the deformed lamina of the trough foresets (Fig. 4.10c). This facies appears to be the most dominant facies of the succession, along with the closely associated next facies, the compound cross-stratified sandstone (Fig. 4.10d). Body geometries, along with the internal structures, indicate such bodies to be channel forms.

Fig. 4.10 Cross-stratifications within Upper Rewa Sandstone: trough cross-stratification (**a**), overturned cross-stratification (**b**) pebbly sandstone showing trough cross-stratification (**c**) and compound cross-stratification (**d**) (hammer length = 38 cm)

4.6.2 Compound Cross-Stratified Sandstone

This facies consists of poorly sorted coarse sandstone bodies, with frequent oversized clasts (Fig. 4.10d). The sandstone bodies are broadly lenticular in flow-perpendicular direction with flat bases and convex-upward tops. In flow-parallel sections, these bodies used to show compound cross-strata, where smaller sets of trough cross-strata are found encased between two successive foresets of larger tabular cross-strata. The larger tabular foresets and smaller trough foresets dip in the same direction, which is also consistent with the overall paleocurrent direction. Always the smaller trough foresets dip more (~25–30°) than the enveloping larger tabular foresets (~5–10°).

Rarely smaller foresets dip in the opposite direction. In such cases, paleocurrent direction revealed from larger foresets appears to be opposite to the overall paleocurrent direction. The oversized pebbles, wherever present, generally defines the larger foreset bases. Some facies units have also preferred pebble concentration on their tops in the form of a discontinuous thin sheet. Some of the smaller foresets within the compound cross-strata shows evidence of overturning. This facies represents the accretionary longitudinal bar forms.

4.6.3 Tabular Cross-Stratified Sandstone

This facies is characterized by tabular cross-stratified medium sand-sized, moderately sorted sandstone bodies, without any oversized clasts (Fig. 4.11a). Such bodies although comprises only a minor fraction of the total succession. Two varieties can be identified within this facies. One subfacies shows broadly lenticular bodies in

Fig. 4.11 Cross-stratifications present within Upper Rewa Sandstone in vertical sections of Chachai falls: tabular cross-stratification (**a**), sigmoidal cross-stratification (**b**), low-angle cross-stratification (**c**), reverse-graded low-angle cross-strata (**d**), bipolar–bimodal cross-stratification (**e**) (scale length in a = 15 cm, pen length = 14 cm, hammer length = 38 cm)

the transverse section with internal tabular foresets dipping conformably with the general paleocurrent direction. This facies is found on top of master erosion surfaces and associated with previous two facies, laterally and vertically. The other subfacies is lenticular in geometry and shows angular relationship with general paleocurrent direction. The former variety represents transverse bars, whereas the latter ones account for the lateral accretions over the longitudinal bar.

4.6.4 Ripple-Laminated Sandstone

This facies is characterized by ripple-laminated medium- to fine-grained moderately sorted, sub-rounded sandstone bodies with planar but patchily preserved body geometry. Rarely some cuspate ripples are observed over the bedding plane. This facies represents the penultimate stages of channel filling, indicating ripple movement over almost filled up channel floors.

4.6.5 Planar-Laminated Sandstone

This facies is characterized by internally planar-laminated, discontinuous, but broadly tabular bodies of very coarse-to-coarse sand-sized materials. It constitutes a comparatively smaller proportion of the total succession, consisting of two subfacies. The first subfacies is comparatively thicker with tabular geometry, and it overlies major erosional boundaries (Fig. 4.11d). It consists of poorly sorted, very coarse sandstone, devoid of pebbles. The other subfacies is made up of the coarsest fraction of the underlying body, with moderate roundness and sorting. The former subfacies represents higher flow regime deposits, forming over the channel bases at the initial stage of channel filling, while the latter subfacies represents the lag deposits formed over the bar tops.

4.6.6 Sigmoidal Cross-Stratified Sandstone

This facies consists of medium-grained sandstone bodies with lenticular geometry in both sections. In both the sections, they show flat base and convex-up top boundary. Internally, the sandstone bodies are made up of sets of sigmoidal cross-strata in longitudinal section (Fig. 4.11b). In transverse section, this facies is characterized by sets of slightly convex-upward cross-laminations. This facies is very much restricted in occurrence. This facies represents linguoid bar forms.

4.6.7 *Reverse-Graded, Gently Dipping Tabular Cross-Stratified Sandstone*

This facies consists of fine-grained, well-sorted sheet sandstone internally characterized by low-angle (<10°) cross-stratification. Strongly asymmetric, straight or broadly sinuous-crested ripples, with height about 2–3 mm only and wavelength as large as 15 cm are locally present on top of the beds or stack of beds. Coarsest grain fraction concentrates along crests of these ripples (Fig. 4.11c). The wavelength: height ratio of these ripples exceeds 15:1. This facies represents translatent strata of eolian origin.

4.6.8 *Bipolar–Bimodal Cross-Stratified Sandstone*

This facies is unique to this section only and scarce in occurrence. It is characterized by alternate layers of oppositely dipping tabular foresets made up of well-sorted, sub-rounded medium sand-sized materials (Fig. 4.11e). This facies represents aeolian linear dunes.

4.6.9 *Bounding Surfaces*

As mentioned earlier, this study location provides an enormous opportunity to study different ranks and orders of fluvial bounding surfaces (Fig. 4.12a, b). Bounding surfaces are defined as 'surfaces of non-deposition or erosion representing time periods of a few minutes to hundreds of thousands of years' (Miall 1988). The hierarchy of the surfaces is expressed by the term 'order' and can be determined following a basic rule that a surface will be one order higher than the highest-order surface it binds (Holbrook 2001). Conceptually such bounding surfaces never cross-cut but older sediments or surfaces may get truncated by a younger surface (Holbrook 2001). The general consensus regarding the assignment of ranks to bounding surfaces is first proposed by Miall (1988), which follows by allotting the zero order to the surfaces bounding the individual laminae and the successively higher denominations for those bounding stratal packages of successively higher genetic orders. The lamina boundaries, cross-set boundaries and coset boundaries are designated as the zero-, first- and second-order surfaces, respectively (Miall 1988; Hjellbakk 1997; Miall and Jones 2003; Holbrook 2001; Holbrook et al. 2006). Third-order bounding surfaces delimit the 'macroforms' (Jackson 1976), like bars and interbar channels. The fourth-order surfaces demarcate channel fills, composed of laterally as well as vertically juxtaposed channel or bar forms. These surfaces are broadly concave upwards or undulatory and laterally extensive. Each channel fill element records filling of major channels during channel avulsion and abandonment (Holbrook 2001). The next order

Fig. 4.12 Fluvial succession showing several orders of bounding surfaces (man for scale = 1.8 m)

surfaces, that is, the fifth-order surfaces are designated as channel-belt boundaries (Fig. 4.12). They are much more extensive in nature, planar in geometry with broad undulations and erosional scours. Such channel belts are made up of both lateral and vertical juxtapositions of several channel fills, both single- and multistoried. Sixth-order surfaces are assigned to the valley-fill boundary status as it binds several fifth-order boundaries, some of which used to terminate against each of them. The segment of the succession in between such two sixth-order surfaces is designated as the valley fill (Holbrook 2001; Holbrook et al. 2006). Valley fill boundaries are considered to be equivalent with the basin-scale unconformity surfaces. Large-scale bounding surfaces are depicted clearly through the excellently preserved continuous outcrops of the area. Several sixth-order bounding surfaces, demarcating the valley fill boundaries, are identified (Fig. 4.12). Total six numbers of valley fills are identified in the study locations. Each valley fill is made up of lateral as well as the vertical juxtaposition of several channel belts, demarcated by fifth-order bounding surfaces (Fig. 4.12). Channel belts, in turn, are made up by vertically as well as laterally juxtaposed channel fills. Each channel fill is composed mostly of channel and bar forms with subordinate other elements.

4.7 Deformation Structures Within Kheinjua Formation in Koldaha

Wide varieties of syn-depositional deformation structures occur within the Koldaha Shale along the northern bank of the River Son, around Koldaha, located south of Chorhat. Despite the wide diversity of these features, they occur at selected stratigraphic intervals and are laterally persistent for many km (Sarkar et al. 1995; Bose et al. 1997, 2001). The dark grey, plane-laminated shale is interbedded with sheets of siltstones and sandstones. The sandstone beds exhibit sole marks at the base and internally they exhibit planar lamination followed by hummocky cross-stratification. The sandstone beds are wave rippled at the top. Very distinctive conglomeratic assemblages and a few isolated granule bodies disrupt the general facies motif of the Koldaha Shale.

Interestingly most of the deformation features are consistently inclined across the Koldaha Shale succession. The most common syn-depositional features include large- and small-scale faults, boudinages, joints, flexures, load casts, convolute lamination, and contorted beds and pillow beds (Fig. 4.13a–g). While large-scale faults and flexures affect multiple beds, the remaining features are bed-confined. Spectacular pillows occur in relatively thicker sandstone beds (70–80 cm thick), while delicate conjugate faults and domino faults affect thin sandstone beds (10–15 cm thick). Contorted beds show wide variations in contortion of internal laminae. Boudinages and extensional faults indicate a roughly northwest–southeast extension. Sarkar et al. (1995) and Bose et al. (1997, 2001) considered seismic origin for many of the features based on the following:

(a) the general stratigraphic selectivity of the deformation structures and considerable lateral continuity of the deformed layers,
(b) consistent orientation of the syn-sedimentary faults in different stratigraphic levels points to a single controlling mechanism like seismic shock,
(c) evidence of growth faults with progressively dying intensity,
(d) large pillows are strongly suggestive of seismic shocks,
(e) anachronistic paleocurrent patterns exhibited by pebbly sandstone and conglomerate beds associated with large-scale extensional faults.

4.8 Tsunamiite Deposits Within Upper Bhander Sandstone in Rampur

A distinctive event bedset is encased within the Upper Bhander Sandstone in Rampur, near Maihar (Fig. 4.14a–e). This part of the Upper Bhander Sandstone is a coastal erg-margin deposit. The event bedset is present at a preferred stratigraphic level and spread over a distance of more than 50 km roughly in strike-parallel direction. The concerned bedset occurs exclusively in five hill-top exposures of the Upper

Fig. 4.13 Soft sediment deformation structures within Koldaha Shale: small-scale domino-style fault (**a**), small-scale conjugate faults (**b**), large-scale fault (**c**), pillowed sandstone bed (**d**), convolute-laminated pillow-like bed (**e**) small-scale load casts (**f**) and slump fold (**g**) (pen length = 14 cm, matchstick length = 4.5 cm, coin diameter = 2.5 cm)

Fig. 4.14 A bed couplet composed of multiple units rapidly thins laterally, draped by clay preserving straight and round crested combined flow ripples underneath themselves and also on the bed surface (**a**), massive, convoluted, pillow-like structures associated with the event bed couplets. Note no evidence of sagging on the undersurface (**b**), tight and imbricated folds, related to the suspected tsunami beds. Note lack of evidence of sagging on the undersurface identifies the plane as that of detachment (**c**), breccia formed within suspected tsunamiite (**d**), flute with multidirectional orientations showing cross-cutting relationship between them under the tsunamiite bed within the coastal eolian deposits. Note the loaded recurved nature of some flutes (**e**) (hammer length = 38 cm, Swiss knife length in a = 9.1 cm)

Bhander Sandstone. In all these five exposures, the bedset is dominantly encased by the components of the erg-margin facies association; wave-rippled sand sheets intervene only locally. There is, however, little doubt that the paleogeography of the concerned bedset had been transitional between the supralittoral zone and the coastal eolian dune field. Close association with warts and adhesion laminae, in particular, suggests a paleogeography within the groundwater capillary fringe zone. The event bedset reported from the lower part of the prograding Upper Bhander Sandstone, central India is sandy and, being Neoproterozoic in age is also free of shelly fossils (cf. Hassler et al. 2000; Pratt 2001). The bedset is present within the coastal Upper Bhander Sandstone erg-margin deposits and they are laterally extensive, tabular in geometry. The internal laminae are poorly developed, and they are less sorted compared to surrounding sediments. The sandstone bed couplets, generally two in number in most locations, comprise the bedset (Fig. 4.14a). Within each couplet, one bed component gives way into the other vertically as well as laterally, always sharply though, their body geometries and internal characters being distinctly different. The erg-margin facies association of ca.25 m thick, overlying a beach deposit, is reworked by storm occasionally. Internally, the erg-margin units are composed of crenulated adhesion laminae, low-angle, planar, inversely graded translatent strata, as well as eolian impact ripples characterized by long wavelength compared to its height. Coarse lags are present along the crests of the ripple. Solitary sets of grainfall–grainflow dune cross-strata occur locally (Bose et al. 1999; Simpson et al. 2004; Sarkar et al. 2011). Towards the top of the succession, the erg facies are more dominant, characterized by large-scale climbing dune cross-strata. The foresets are often inversely graded. Occasionally plane beds, pinstripe lamination and impact ripples are also present. The sandstone is characteristically well sorted. Within the bedset, there are two components. In one component, they rest on pronounced concave-up erosional surfaces and have rapidly wedging geometries, up to 45 cm in thickness and thin rapidly.

They are up to 40 cm in thickness, but thin rapidly landward as they fill and spill over, onlapping the seaward-sloping basal scours (Fig. 4.14a). Each bed of the sandstones overlying these scours (maximum depth 30 cm) is composed of multiple vertically stacked subunits with supercritically climbing ripples and a thin mud drape. Wave-cum current, combined flow ripple with rounded crest, occasionally bifurcating and strongly asymmetric are well preserved under the clay drapes. The other components of this sandstone have planar basal erosion surfaces except for the minor sole features. Their body geometry is tabular, and the maximum thickness of individual beds is 28 cm. The beds are generally massive or normally graded. Small-scale folds, convolutes and pillow-like structures are associated with them (Fig. 4.14b). The folds may be tight, imbricated (Fig. 4.14c), and locally recumbent, yet their tops are often not truncated and surfaces underlying them, being planar, bear no evidence of sagging (Fig. 4.14b, c). Exhumed slump scars, in rows, are often clearly associated with them. In the same direction roughly, the tight folds locally pass into breccias first (Fig. 4.14d), and then into the massive sandstone. Soles of the sandy part of the beds bear a few gutter casts and numerous flute casts, which are often recurved (Fig. 4.14e). The most significant aspects of the flutes are their multiple generations as attested by their mutual cross-cutting relationship, the loaded nature of many (Fig. 4.14e; Mastalerz 1995), and distinct temporal diversions in orientation. Considering the suite of sedimentary features, structural and textural components of the concerned bedsets, it strongly suggests tumultuous or 'convulsive deposition' (Clifton 1988; Sarkar et al. 2011) and erosion and reflects a depositional condition drastically different from that of all other beds of the Upper Bhander Sandstone. Flow and depositional dynamics inferred from this bedset seem to be more in accord with a tsunami interpretation than any other high energy natural event that is paleogeographically compatible.

Coordinates of Places Mentioned in this Chapter

Bhadanpur	24°9′35.47″N; 80°49′2.51″E
Kymore	24°3′35.18″N; 80°36′27.91″E
Maihar	24°16′9.89″N; 80°45′23.94″E
Amiliya	24°15′28.89″N; 80°48′18.29″E
Girgita	24°15′28.89″N; 80°48′18.29″E
Dolni	24°18′10.72″N; 80°48′11.49″E
Chachai	24°47′47.18″N; 81°18′10.11″E
Koldaha	24°24′27.7″N; 81°40′31.4″E
Beta	24°25′41.22″N; 80°46′51.86″E
Rampur	24°16′9.84″N; 80°41′13.52″E

References

Awramik SM (1971) Precambrian columnar stromatolite diversity: reflection of metazoan appearance. Science 174:825–827

Awramik SM, Margulis L (1974) Definition of stromatolite. In: Walter E (ed) Stromatolite Newsletter vol 2, no 5

Banerjee S, Bhattacharya SK, Sarkar S (2007) Carbon and oxygen isotopic variations in peritidal stromatolite cycles, Paleoproterozoic Kajrahat Limestone, Vindhyan Basin of central India. J Asian Earth Sci 29:823–831

Beukes NJ, Lowe DR (1989) Environmental control on diverse stromatolite morphologies in the 3000 Myr Pongola Supergroup, South Africa. Sedimentol 36:383–397

Bose PK, Chakraborty PP (1994) Marine to fluvial transition: Proterozoic Upper Rewa Sandstone, Maihar, India. Sed Geol 89:285–302

Bose PK, Sarkar S (1991) Basinal autoclastic mass flow regime in the Precambrian Chanda Limestone Formation, Adilabad, India. Sed Geol 73:299–315

Bose PK, Banerjee S, Sarkar S (1997) Slope-controlled seismic deformation and tectonic framework of deposition of Koldaha Shale, India. Tectonophys 269:151–169

Bose PK, Chakraborty S, Sarkar S (1999) Recognition of ancient aeolian longitudinal dunes: a case study from the Upper Bhander Sandstone, Son Valley, India. J Sed Res 69:86–95

Bose PK, Sarkar S, Chakraborty S, Banerjee S (2001) Overview of the Meso-to Neoproterozoic evolution of the Vindhyan basin, central India. Sed Geol 141:395–419

Bose PK, Eriksson PG, Sarkar S, Wright P, Samanta P, Mukhopadhyay S, Mandal S, Banerjee S, Altermann W (2012) Sedimentation patterns during the Precambrian: a unique record? Mar Petrol Geol 33:34–68

Bottjer JW, Hagadorn, JW (2007) Mat growth features. In: Schieber J, Bose PK, Eriksson PG Banerjee S, Sarkar S, Catuneanu O, Altermann W (eds) An Atlas of Microbial Mat Features Preserved within the Clastic Rock Record, Elsevier, pp 53–71

Bouougri E, Porada H (2007) Mat related features from the terminal Ediacaran Nudaus Formation, Nama Group, Nambia, Schieber. In: Schieber J, Bose PK, Eriksson PG, Banerjee S, Sarkar S, Altermann W, Catuneanu O (eds) Atlas of Microbial Mat Features Preserved within the Siliciclastic Rock Record. Elsevier, Amsterdam, pp 214–221

Chakraborty PP (2004) Facies architecture and sequence development in a Neoproterozoic carbonate ramp: Lakheri Limestone Member, Vindhyan Supergroup, central India. Precambr Res 132:29–53

Chakraborty PP (2011) Slides, soft-sediment deformations, and mass flows from Proterozoic Lakheri Limestone Formation, Vindhyan Supergroup, central India, and their implications towards basin tectonics. Facies 57:331–349

Clifton HE (1988) Sedimentologic relevance of convulsive geologic events. In: Clifton HE (ed) Sedimentologic consequences of convulsive geologic events. Geological Society of America Special Papers, vol 229, pp 1–5

Enos P (1977) Flow regimes in debris flow. Sedimentology 24:133–142

Frey SE, Gingras MK, Dashtgard SE (2009) Experimental studies of gas-escape and water-escape structures; mechanisms and morphologies. J Sed Res 79:808–816

Hassler SW, Robey HF, Simonson BM (2000) Bedforms produced by impact generated tsunami, 2.6 Ga Hammersley basin, Western Australia. Sed Geol 135:283–294

Hjellbakk A (1997) Facies and fluvial architecture of a high-energy braided river: the upper Proterozoic Seglodden Member, Varanger Peninsula, northern Norway. Sed Geol 114:131–161

Hofmann HJ (1973) Stromatolites: characteristics and utility. Ear Sci Rev 9:339–373

Holbrook JM (2001) Origin, genetic interrelationships, and stratigraphy over the continuum of fluvial channel-form bounding surfaces: an illustration from middle Cretaceous strata, southeastern Colorado. Sed Geol 124:202–246

Holbrook JM, Scott RW, Oboh-Ikuenobe FE (2006) Base-level buffers and buttresses: a model for upstream versus downstream control on fluvial geometry and architecture within sequences. J Sed Res 76:162–174

Jackson RG (1976) Hierarchical attributes and a unifying model of bed forms composed of cohesionless material and produced by shearing flow. Geol Soc Am Bull 86:1523–1533

Jüngst H (1934) Zur geologischen Bedeutung der Synärese. Geol Rundschau 15:312–325

Kalkowsky E (1908) Oolith und Stromatolith in Norddeutschen Buntsandstein. Zeitschrift der Deutschen Gesellschaft für Geowissenschaften 60:68–125

Mastalerz K (1995) Deposits of high-density turbidity currents on fan-delta slopes: an example from the upper Visean Szczawno Formation, Intrasudetic basin, Poland. Sed Geol 98:121–146

Miall AD (1988) Reservoir heterogeneities in fluvial sandstone: lessons from outcrop studies. Am Ass Petrol Geol Bull 72:682–687

Miall AD, Jones BG (2003) Fluvial architecture of the Hawkesbury Sandstone (Triassic), near Sydney, Australia. J Sed Res 73:531–545

Plummer PS, Gostin VA (1981) Shrinkage cracks—desiccation or synaeresis? J Sed Pet 51:1147–1156

Pratt BR (2001) Oceanography, bathymetry and syndepositional tectonics of a Precambrian intracratonic basin: integrating sediments, storms, earthquakes and tsunamis in the Belt Supergroup (Helena Formation, ca. 1.45 Ga), western North America. Sed Geol 141/142: 371–394

Riding R (2000) Microbial carbonates: the geological record of calcified bacterial-algal mats and biofilms. Sedimentol 47(Suppl. 1):179–214

Sarangi S, Gopalan K, Kumar S (2004) Pb–Pb age of earliest megascopic, eukaryotic alga bearing Rohtas Formation, Vindhyan Supergroup, India: implications for Precambrian atmospheric oxygen evolution. Precam Res 132:107–121

Sarkar S, Bose PK (1992) Variations in late Proterozoic stromatolite forms over a transition from basin plain to nearshore subtidal. Precam Res 56:149– 157

Sarkar S, Banerjee S, Chakraborty S (1995) Synsedimentary seismic signatures in the Mesoproterozoic Koldaha Shale, Kheinjua Formation, central India. Indian J Earth Sci 22:158–164

Sarkar S, Eriksson PG, Chakraborty S (2004) Epeiric sea formation on Neoproterozoic supercontinent break-up: a distinctive signature in coastal storm bed amalgamation. Gond Res 7:313–322

Sarkar S, Bose PK, Eriksson PG (2011) Neoproterozoic tsunamiite: Upper Bhander Sandstone, central India. Sed Geol 238:181–190

Sarkar S, Samanta P, Mukhopadhyay S, Bose PK (2012) Stratigraphic architecture of the sonia fluvial interval, India in its Precambrian context. Precam Res 214:210–226

Sarkar S, Choudhuri A, Banerjee S, van Loon (Tom) AJ, Bose PK (2014) Seismic and non-seismic soft-sediment deformation structures in the Proterozoic Bhander Limestone, central India. Geologos 20:79–93

Schieber J (2004) Microbial mats in the siliciclastic rock record: a summary of the diagnostic features. In: Eriksson PG, Altermann W, Nelson DR, Muller WU, Catuneanu O (eds) The Precambrian Earth: Tempos and Events. Elsevier, Amsterdam, pp 663–673

Simpson EL, Alkmim FF, Bose PK, Bumby AJ, Eriksson KA, Eriksson PG, Martins Neto MA, Middleton LT, Rainbird RH (2004) Sedimentary dynamics in Precambrian aeolianites. In: Eriksson PG, Altermann W, Nelson DR, Mueller W, Catuneanu O (eds) The Precambrian Earth: Tempos and Events. Elsevier, Amsterdam, pp 675–677

Tosti F, Riding R (2017) Fine-grained agglutinated elongate columnar stromatolites: Tieling Formation, ca 1420 Ma, North China. Sedimentol 64:871–902

Chapter 5
Microbial Mat Structures Formed Within Siliciclastics

5.1 Introduction

Microbiota appeared on Earth before 3.8 Ma, and they flourished during the Proterozoic time (Schopf 1999). Subsequently, they became virtually confined to stressful environments where environmental conditions remained too harsh for most grazing and burrowing organisms (Hagadorn and Bottjer 1999; Schieber 1999, 2004). The physical, chemical and biotic ambiance of any sedimentary environment undergoes many changes in the presence of microbiota. The recognition of microbiota or microbial mat-related features within sedimentary rocks has enormous potential for understanding the evolution of early life on Earth (Sarkar et al. 2016). In the Precambrian era, microbial mats probably colonized on most of the sedimentary surfaces both in carbonate platforms and in siliciclastic settings (e.g. Schieber 1999; Schieber et al. 2007; Banerjee and Jeevankumar 2005; Banerjee et al. 2014). They usually leave detectable signature within the sedimentary pile, especially at the sediment–water interface. Despite their numerical supremacy, the microbiota has a poor geological record in comparison with fossil records of macrobiota. Nonetheless, their colonies may leave unique sedimentary structures, contemporary biotic and abiotic processes subsequently destroy them. Three-dimensional convex-up columnar stromatolite is one such structure that is long known from carbonate rocks from all over the world (e.g. Kalkowsky 1908; Walter 1972; Grotzinger 1990; Sarkar and Bose 1992; Choudhuri et al. 2016). Originated from the Greek word 'stroma' meaning mattress, bed or stratum, and 'lithos' meaning rocks, stromatolite represents a convex-up layered sedimentary structure commonly found in carbonate formations of pre-Devonian age (Grotzinger 1990; Shen and Webb 2004; Flügel 2013; Noffke and Awramik 2013). Modern stromatolites are archetypal colonies of bacteria, especially cyanobacteria (Awramik and Margulis 1974; Burne and Moore 1987; Noffke and Awramik 2013; Lan and Chen 2013). It is, therefore, reasonable to assume microbial origin for some or most of the ancient stromatolites as well.

© Springer Nature Singapore Pte Ltd. 2020
S. Sarkar and S. Banerjee, *A Synthesis of Depositional Sequence
of the Proterozoic Vindhyan Supergroup in Son Valley*, Springer Geology,
https://doi.org/10.1007/978-981-32-9551-3_5

Several workers reported the occurrence of stromatolites in the Vindhyan Supergroup (Table 1.3). Banerjee et al. (2007) identified a regular and systematic change in stromatolite size variations within the Kajrahat Limestone. Several shallowing-upward stromatolite cycles comprising larger stromatolites at the base followed upwards by smaller stromatolites, and microbial laminites at the top represent shallowing of the water level leading to subaerial exposure (Banerjee et al. 2007). Systematic variations in size and morphology of stromatolites are also well documented from the Lakheri Limestone in Upper Vindhyan (Sarkar et al. 1996).

A array of bedding plane structures with similar microbial mat implications has been reported from terrigenous siliciclastic rocks (e.g. Hagadorn and Bottjer 1997; Schieber 1998, 1999; Gehling 1999; Pflüger 1999; Seilacher 1999; Eriksson et al. 2000; Gerdes et al. 2000; Noffke 2000, 2010; Noffke et al. 2001, 2002, 2003a; Bouougri and Porada 2002; Sarkar et al. 2004, 2006, 2008; Parizot et al. 2005; Schieber et al. 2007). Direct signatures of microbiota are either rare or very localized in terrigenous sedimentary rocks. Proxy structures, however, resulting from the interaction between microbiota and the sediment, such as mat-induced sediment binding, grain agglutination and chemical compartmentalization of the deposit, are common in shallow marine sandstones and offshore shales. To a large extent, these proxy structures owe their formation to trapping, binding, baffling and biostabilization by the extracellular polymer substances (EPS) secreted by cyanobacteria and other microorganisms (Decho 1990, 2000). EPS makes sand and watery mud cohesive. Granular and non-cohesive sand responds differently to stress, often behaving more like mud, and forms a group of features. These features, such as desiccation cracks, sand–curls, pebble-sized, flat or rolled-up sand fragments are generally not expected within sandstones. These two-dimensional microbial proxy structures are commonly known as mat-induced sedimentary structures (MISS; Noffke 2010). Some of these structures are mat-diagnostic because they owe their preservation to microbial mat growth, though their origin does not (Sarkar et al. 2011). Hence, they are also termed as mat-related structures (MRS; Schieber et al. 2007).

5.2 Microbial Mat-Related Structures (MRS)

Precambrian siliciclastics remained unexplored for microbial mat record for many years. However, since the last decade several workers reported microbial mat-related structures in Precambrian siliciclastics within the Vindhyan Supergroup (Sarkar et al. 2004, 2005, 2006, 2008, 2011, 2014, 2018; Banerjee and Jeevankumar 2005; Banerjee et al. 2006, 2010, 2014; Sur et al. 2006; Sarkar and Banerjee 2007; Bose et al. 2007; Schieber et al. 2007). Chorhat Sandstone and Koldaha Shale members of the Kheinjua Formation display a spectacular variation of MRS. Sirbu Shale and Upper Bhander Sandstone within the Bhander Formation also exhibit wide-ranging microbial mat structures. Following discussion provides a brief description of the microbial mat structures in the Vindhyan Supergroup.

5.2.1 *MRS in Kheinjua Formation*

5.2.1.1 Cracks

Cracks, millimetre to a centimetre in width, are common in occurrence on the bedding surfaces of the well-sorted Chorhat Sandstone. The geometry of the cracks may vary from rectangular-, polygonal-, circular-, sinusoidal- to spindle-shaped (Fig. 5.1a–f). They may form a network or may run parallel to each other. In the networks, they often cut across each other (Fig. 5.1a, f). They are invariably steep-sided and may occur selectively along crests or troughs of wave ripples (Fig. 5.1d, e).

Generation of cracks on sand suggests the influence of mat that contributes cohesion to the granular, non-cohesive sediment. Although the mat has now disappeared

Fig. 5.1 Cracks of various shapes on sandstone beds: rectangular (**a**), polygonal (**b, c**), circular (**d**), sinusoidal crack (**e**) and spindle-shaped (**f**) (Swiss knife length = 9.1 cm)

on degeneration and decomposition, the cracks bear a proxy record of the pristine mat. Desiccation might have been responsible for cracking of the mat-infested sediment surface (Schieber 2004; Schieber et al. 2007). However, those cross-cutting cracks possibly form by syneresis (Fig. 5.1a, f). The occurrence of the majority of cracks in sediments deposited below the sea surface corroborates the syneresis origin (Facies A and lower part of Facies B unit, see Table 2.5).

Confinement of cracks, generally sinusoidal in the plan, often within ripple troughs is possible because of thicker mat present within troughs. The sinusoidal geometry of cracks relates to micro-scale variation in mat topography (Pflüger 1999; Schieber 2004; Parizot et al. 2005). In certain instances, these trough-confined cracks are short, straight and oriented at an acute angle to the trough axis. Schieber (2004) and Parizot et al. (2005) attribute similar cracks to desiccation of mats. This structure is widely known as '*Manchuriophycus*'. Whether generated on exposed surface or underwater, they are now widely accepted as mat features (Gehling 1999; Parizot et al. 2005), the occasional invocation of trace fossil origin notwithstanding (e.g. Kulkarni and Borkar 1996).

Spindle-shaped cracks along ripple crests owe their origin to shrinkage of very thin surficial sediment on top of the rippled sand. Since these cracks are unknown in modern mats, it is difficult to say whether they form at the sediment–air interface, sediment–water interface or in shallow subsurface (Parizot et al. 2005). The tensile shear needed for their formation might have been induced by the desiccation or by sediment over-burden (Sarkar et al. 2004). Bevelled edges of the surface layer created by the cracks may give a false impression of abnormal sharpness of the ripple crests formed under combined wave and current actions.

5.2.1.2 Ridges

Ridges with heights up to 0.5 cm, showing similar geometric variation as cracks, are also common on sandstone bed surfaces (Fig. 5.2a–d). The ridges are generally wider than the cracks. Unlike cracks, they generally have a rounded bottom in profile. While forming a network, these ridges may cross-cut each other. Small-scale antiform structures originated through buckling, doming and rupturing of microbial mats, are known as petee structures. These are reported from modern settings (Reineck et al. 1990; Gerdes et al. 1993, 1994) as well as from the rock record (Gehling 1999) and have been attributed to winds, currents and gas formation as well as intermittent drying of the microbial mat layer (Schieber 1999, 2004). Perhaps, a more likely explanation of these linear ridges would be the deformation of a syneresis crack in response to the flowage of the cohesive mat layer converging from both sides of the depression under confining pressure. The presence of median furrows in many such ridges depicts the meeting of two flanks of the cracks, thus corroborating this inference (Fig. 5.2a, b). The crack fills stand out on bed surfaces as the sandstones surrounding them erode on exhumation. These ridges have the potential for misguided identification as burrows.

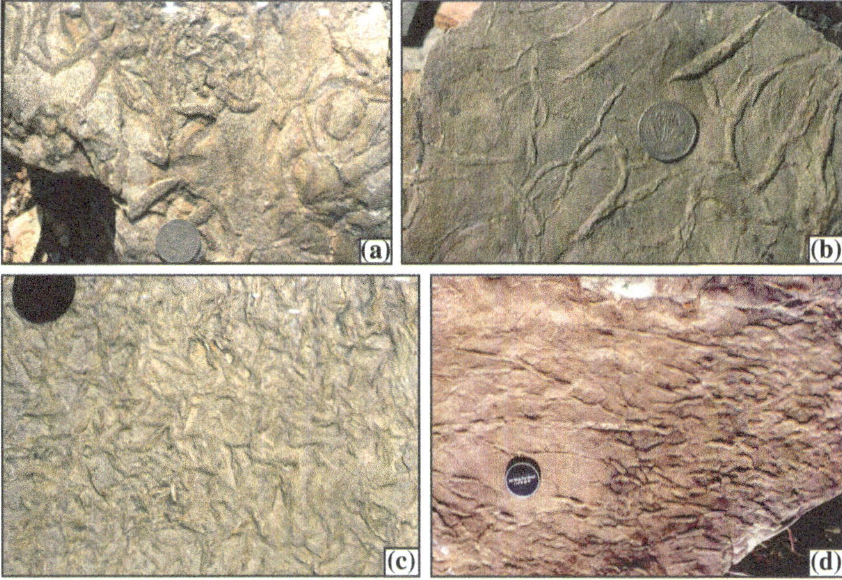

Fig. 5.2 Photographs showing different types of petee ridges on sandstone beds (**a–d**) (lens cap diameter = 4.1 cm, coin diameter = 2.5 cm)

5.2.1.3 Wrinkle Structures

Wrinkle structures are common throughout the Chorhat Sandstone (Fig. 5.3a, b). These may appear as minute ridges, generally with a crinkled appearance. They may be parallel to each other (Kinneyia ripples) or may form a network. In networks, they may cross-cut each other (Fig. 5.4b) or may share the same crests at their meeting points (Fig. 5.4 a, b).

Wrinkle structures on sandstone bed surfaces represent growth of microbial mat, especially where mud is absent. Loading and dewatering during the burial of micro-

Fig. 5.3 Wrinkle structures on sandstone bed surfaces (**a, b**) (lens cap diameter = 4.1 cm, coin diameter = 2.5 cm)

Fig. 5.4 Kinneyia ripples on crests of ripples on a sandstone bed (**a**), Kinneyia ripples on a sandstone bed (**b**)

bial layers explain their origin (Noffke et al. 2003a). Margins between very shallow load depressions may, indeed, appear as wrinkles. In modern intertidal settings, wrinkle marks form on the mat-covered surface. Gas buildup related to the decaying mat can generate wrinkle structures commonly with steep-dipping troughs (Pflüger 1999; Gerdes et al. 2000). A gentle wave or current shear may also form wrinkle structures on the mat (Hagadorn and Bottjer 1997; Bouougri and Porada 2002; Banerjee and Jeevankumar 2005; Banerjee et al. 2010, 2014). Wrinkle structures with unidirectional asymmetry in profile occurring within the Chorhat Sandstone might have developed under shear of gentle wave or current on a microbially mediated cohesive sand surface. The two sets of cross-cutting wrinkles within the Chorhat Sandstone correspond to variably directed flow shears (cf. Banerjee and Jeevankumar 2005).

5.2.1.4 Ruptured Domes and Sand Bulges

Domal structures with a crater-like depression and of different diameters (up to 5 cm) are locally common on sandstone bedding surfaces (Fig. 5.5). Some of them have radiating cracks around them and are named as '*Astropolithon*' described by Pflüger

Fig. 5.5 Small domal structure on a sandstone bed (coin diameter = 1.9 cm)

Fig. 5.6 Small-scale unruptured domes on a sandstone bed

(1999). Expulsion of fluid, mainly gas generated by decay at the bottom of microbial mat cover, is an acceptable explanation for these structures. The radial cracks around them corroborate mat-induced cohesiveness of the sand. Unruptured domes, about half a centimetre high and about a centimetre in diameter are locally common within the Chorhat Sandstone (Fig. 5.6). They may coexist with the ruptured domes on the same bed surface. Ruptured domes and sand bulges are formed by the upward escape of gases through the mat-covered surface (Schieber et al. 2007).

5.2.1.5 Patchy Ripples, Erosional and Depositional

The patchy occurrence of discrete sets of ripples is a common feature throughout the Chorhat Sandstone (Fig. 5.7a, b). These are broadly of two types. In the first case, the preservation of a single train of ripples is patchy, margins of the ripple patches are jagged and stepped and the bed surfaces between the patches are planed off (Fig. 5.7a). In the second case, a first-generation ripple is reworked by a later process, but in a few spots (Fig. 5.7b).

In the first case, the critical question is how the ripple patches survive after the onslaught of a strong flow shear that generally planed off the bed surface. A protective mat cover on the ripple patches has to be invoked. Possibly the mat cover had been widespread but was torn away from most parts by a strong current or wave. Ripples were preserved under the residual mat. Jagged and stepped margins of the patches strongly corroborate this contention. In the second case, the first-generation ripple was reworked subsequently, but in patches; a mat cover must have been present where it escaped reworking.

5.2.1.6 Load Casts

Many sandstone bed surfaces at the lower part of the Chorhat Sandstone bear closely spaced steep-sided depressions (depth < 1 cm), almost circular in cross section and diameter ranging from 0.5 to 3.5 mm, more or less uniform on a single bed surface (Fig. 5.8a, b). On rippled bed surfaces, the depressions have preferred occurrence

Fig. 5.7 Patchy ripples on bed surfaces with planed off surface in between (**a**) and reworking of first-generation ripples by another set in patches (**b**) (pen length = 14 cm, Swiss knife length = 9.1 cm)

within the ripple troughs (Fig. 5.8c). Several interfaces between amalgamated sandstone beds when prized open show the depressions on top of the underlying bed and their casts at the sole of the overlying bed. No mud exists at the interfaces, not even within the depressions.

The depressions and their mound-like casts at the sole of the overlying bed form by loading. The underlying sandstone bed must have turned thixotropic because of the loading. The bed lying above must have been deposited rapidly, possibly from the storm-induced flow. The deformation involved is syn-sedimentary. Perhaps only gelatinous microbial mats could turn granular sand thixotropic at the top of the underlying sandstone bed. The preferred occurrence of the depressions along ripple troughs can then be readily explained by the preferred growth of mat within the troughs.

Fig. 5.8   Load casts on rippled sandstone bedding surfaces (**a–c**) (Swiss knife length = 9.1 cm)

5.2.1.7   Sand Chips

Some bed surfaces in Facies A are mantled by a lag of sand chips (Fig. 5.9). The chips are spherical, ellipsoidal, triangular, crescent-shaped, or sometimes do not show any definite geometry. Crescent-shaped chips appear to be deformed. The chips are, however, generally rounded at their edges, up to 5 cm long. The sand chips consist of well-sorted and well-rounded grains. Sand chip-bearing intraformational conglomerates have been reported from all ages (Menzies 1990; Pflüger and Greese 1996). Early carbonate cement or mineral cement formation between grains before reworking can explain the generation of clasts from granular sand. The Chorhat Sandstone is devoid of carbonate or evaporite cement that can be attributed to early diagenesis. Most common cements include iron oxide and silica that succeeds pressure welding

Fig. 5.9 Sand chips on sandstone bedding surface (Swiss knife length = 9.1 cm)

of sand grains. Deformation indicates a flexible response of the sand clasts and therefore implies the presence of an elastic binding agent, such as microbial mat surrounding the sand particles (Pflüger and Greese 1996). Occasional storm-induced flows were presumably responsible for the erosion of the mat-infested sand layers.

5.2.1.8 Silt Curls

Curls of mud-free coarse siltstone up to 18 cm long are present, albeit rarely, on bed surfaces at the lower part of the Chorhat Sandstone (Fig. 5.10a, b). Their diameter varies from 2 to 4 cm. The silt curls on a single bed surface are oriented parallel to each other (Fig. 5.10a, b).

The silt curls in the Chorhat Sandstone are virtually similar in form with mud curls that are often generated on desiccation. The mutually parallel orientation of the curls suggests reworking by strong flows. It is, however, intriguing how such curls form in granular, non-cohesive silt, and how these curls could sustain reworking by a strong flow. The curled sediments must have acquired a very high degree of cohesiveness, presumably because of microbial mat growth. Similar sandy or silty mat curls formed by desiccation are common in modern tidal flats (Reineck et al. 1990; Witkowski 1990) and their ancient equivalents are reported from a few shallow marine successions (Noffke et al. 1997; Demicco and Hardie 1994; Schieber et al. 2007). Eriksson

Fig. 5.10 Silt curls on sandstone bed surfaces (**a** and **b**) (coin diameter = 2.2 cm)

et al. (2000) reported similar parallelly oriented sand roll-up structures from 1.8 Ga eolian interdune deposit of Waterberg Group, South Africa.

5.2.1.9 Palimpsest Ripples

These are ripple forms replicated from the underlying bed (Fig. 5.11). For such replication, the overlying bed has to be very thin. This structure is commonly found at the lower part of the Chorhat Sandstone with an occasional amalgamation of storm beds. Palimpsest ripples (Pflüger and Sarkar 1996) wrongly depict the hydrodynamic condition of deposition of the bed on which they are found. They carry the memory of hydrodynamic conditions that prevailed during the deposition of the underlying bed. Such replication is attributed to trapping of particles by filaments of microbial

Fig. 5.11 Palimpsest ripples on bedding surfaces (Swiss knife length = 9.1 cm)

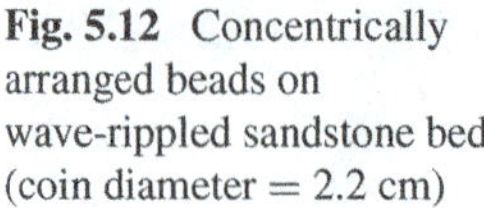

Fig. 5.12 Concentrically arranged beads on wave-rippled sandstone bed (coin diameter = 2.2 cm)

mat, grown uniformly over the ripples formed earlier (Pflüger and Sarkar 1996). A combination of a very low rate of sedimentation and complete cessation of erosion is an essential prerequisite for the generation of palimpsest ripple (Noffke 2000). Degree of replication may, however, vary. When ripples on the underlying bed are replicated well, the structure is called 'transparent', and when thicker mat growth within ripple troughs renders the ripple geometry unrecognizable, the structure is called 'non-transparent' (Fig. 5.12; Noffke et al. 2003a). Thicker growth of mat within depressions tending to remove relief variation on the depositional sub-stratum thus leads to 'levelling' (Noffke et al. 2001).

Though both MRS/MISS and stromatolites result by the interaction between benthic microbial communities and the sediments through biostabilization, binding, baffling and trapping mechanisms. MRS, being mainly a surface phenomenon, remains planar in nature and lacks the considerable height of stacked layers, unlike the stromatolites. Therefore, MRS are two dimensional in nature, whereas stromatolites are three dimensional as they gain significant thickness over time through the above-mentioned processes along with repetitive carbonate precipitation and cementation within the biofilm stacked one above another (Noffke and Awramik 2013). Both MRS and stromatolites are close relatives of the microbial mat, but so far their spatial distribution is compartmentalized in siliciclastic/evaporitic environments and carbonate platforms, respectively.

5.2.1.10 Concentrically Arranged Beads

Upper bedding surface of a wave-rippled sandstone displays small elongated beads of sand arranged in close concentric rims or spires (Fig. 5.12). Individual beads are spindle-shaped and may be up to a few mm long. The beads are consistent in length, width and height. The overall disc-shaped structure has a diameter of ca. 9 cm. The feature resembles the impression of a multicellular organism. However, similar concentric arrangements of tiny beads are formed on modern microbial mats by the ring-like propagation of growth fronts of chemotactic bacteria (Sarkar and Banerjee 2007).

5.2.2 Features in Sirbu Shale

Soles of shelf storm sandstone beds exhibit many varieties of microbial mat structures (Sarkar and Banerjee 2007; Banerjee et al. 2010, 2014). However, similar features may exist as positive hypo-reliefs on the bedding surfaces. The majority of these features are present within the distal shelf originated facies within the Sirbu Shale. Wrinkle structures and related microbial mat structures are locally abundant within the Sirbu Shale. Some of these features resembling Ediacaran fossils are discussed below.

5.2.2.1 Arumberia

Sandstone bed surfaces often exhibit mm-scale radial grooves alternating with mm-scale ridges (up to 1.5 mm high) preferably on the crests of the ripple and form peculiar circular to semi-circular patterns in places (Fig. 5.13a). Similar features were considered initially as *Arumberia*, a cup-shaped animal trace of coelenterate, preserved as positive epi-reliefs (Glaessner and Walter 1975). The growth of microbial mat may also form *Arumberia*-type feature (Bland 1984; Mcilroy and Walter 1997; Noffke 2007). Miniscule ridges possibly form by the action of currents on soft and cohesive microbial mat cover (Mcilroy and Walter 1997; Mcilroy et al. 2005).

Fig. 5.13 Alternate ridges and grooves (*Arumberia*) on a sandstone bed (**a**), wrinkled masses with circular outlines at the sole of a sandstone bed (**b**), sub-rounded impression showing wrinkled mass at the bottom of a sandstone bed (**c**) (pen length = 14 cm, matchstick length = 4.5 cm)

5.2.2.2   Disc-Shaped Microbial Colonies

The soles of fine-grained storm sandstone beds bearing distinct prod marks may often exhibit roughly circular, wrinkled masses preserved as positive hyporelief features (Fig. 5.13b). The diameter of the discrete circular masses ranges from 2.7 to 3.7 cm. The wrinkled masses appear very similar to Ediacaran fossil *Nimbia* (Fedonkin 1980; see also De 2003, 2006). However, the Ediacaran origin of the feature is ruled out based on radiometric and absence of characteristic features of Ediacaran fossils. The feature may either represent impressions of torn pieces of microbial mats or impressions of a disc-shaped microbial colony (Sarkar and Banerjee 2007; Banerjee 2012; Banerjee et al. 2014).

The lower bedding surfaces of storm sandstone beds often exhibit positive epirelief features with a discrete ovate outline marked by sharp but partly eroded peripheral ridges (Fig. 5.13c). The length of these sub-rounded features varies from 1.5 to 2.1 cm. Internally, these features show wrinkles consisting of tiny ridges arranged in a concentric fashion. Although the feature looks like the Ediacaran fossil *Kaisalia* (cf. Fedonkin 1984), it is comparable to the disc-shaped microbial colony (cf., Grazhdankin and Gerdes 2007). The latter interpretation is more likely considering the > 600 Ma age of the Bhander Formation (see Chap. 1).

5.3   Implications of Microbial Mat Structures

MRS has recently been studied in great detail within the siliciclastic rocks (e.g. Banerjee and Jeevankumar 2005; Banerjee et al. 2010, 2014; Eriksson et al. 2000, 2010; Gehling 1999; Gerdes et al. 1993, 1994, 2000; Hagadorn and Bottjer 1999; Noffke et al. 2001, 2003b, 2013; Pflüger 1999; Prave 2002; Sarkar et al. 2004, 2006, 2008, 2011, 2014; Schieber 1998, 1999; Schieber et al. 2007; Banerjee 2012, 2013) but reports of such features from ancient carbonate deposits are rare and limited in number, although MRS is known in modern counterparts (e.g. Bose and Chafetz 2011). Nevertheless, Luo et al. (2013) and Xiaoying et al. (2008) reported wrinkle structure, a kind of MRS from carbonate formations. Sarkar et al. (2016, 2018) considerably expanded the list of 2D carbonate MRS from the Rohtas Limestone and Bhander Limestone of the Vindhyan Supergroup. The rarity of MRS in ancient carbonate formations may primarily be related to early cementation, and the said structures, in almost all cases, involve micro-scale deformation. Besides being two-dimensional in nature, MRS is far less conspicuous than stromatolites and hence might have generally been overlooked in ancient carbonate formations. The relative scarcity of bedding plane exposures because of pervasive early cementation of sediments may also be responsible for rare findings of MRS in these formations. Even when bedding planes are exposed, intense weathering renders recognition of MRS difficult. Some mat-diagnostic structures arising from trapping, binding and stabilization of sedimentary particles have the connotation of clastic sedimentation, effectively denied in many carbonate depositional systems. Lastly, the majority of MRS typically develops in terrigenous siliciclastic sediments that involve soft-sediment deformation. The implied delayed cementation may be attributed to acidic composition of the main components of the EPS of microbial mats. Besides, the thinness of microbial mats in restricted environments severely delimits the activity of sulphur-reducing bacteria, the main promoter of $CaCO_3$ precipitation (Sarkar et al. 2016).

MRS identified within the different siliciclastics and carbonate formation of the Vindhyan Supergroup indicate the profused microbial mat growth on sediment substrate during the Precambrian. The microbiota had a dominant role in sequence building pattern during the Precambrian (Sarkar et al. 2005, 2014, 2016; Banerjee and Jeevankumar 2005; Chakraborty et al. 2012). The siliciclastic successions exhibit contrasting trends of sequence building in comparison to those in Phanerozoic sedimentary basins. General lack of well-developed transgressive systems tracts (TSTs) is common within the Proterozoic clastic system. Instead, they show the dominance of stacked prograding and aggrading 'normal regressive' systems' tracts (Sarkar et al. 2005, 2016; Catuneanu 2006). Within the Vindhyan sequences, the aggradation of normal regressive deposit relates to the prolific growth of microbial mats in spite of low sedimentation (Catuneanu 2006). The lush growth of microbial mats prevented erosion of microbially bound sediments (Sarkar et al. 2005), and therefore stacked packages of normal regressive systems tract should be the character of the sequence stratigraphic architecture of the Precambrian successions instead of transgressive–regressive systems tract, as usually observed in the Phanerozoic successions (Sarkar et al. 2005, 2014; Banerjee and Jeevankumar 2005; Chakraborty et al. 2012).

References

Awramik SM, Margulis L (1974) Definition of stromatolite. In: Walter E (ed) Stromatolite newsletter, vol 2, p 5

Banerjee S (2012) Discoidal microbial colonies. Int J Earth Sci 101:1343

Banerjee S (2013) Microbial mat-related peculiar tunnels and cracks from modern supratidal environment, Gulf of Cambay, India. Int J Earth Sci 102:2223

Banerjee S, Jeevankumar S (2005) Microbially originated wrinkle structures on sandstones and their stratigraphic context: Paleoproterozoic Koldaha Shale, Central India. Sed Geol 176:211–224

Banerjee S, Bhattacharya SK, Sarkar S (2006) Carbon and oxygen isotope compositions of the carbonate facies in the Vindhyan Supergroup, Central India. J Earth Sys Sci 115:113–134

Banerjee S, Bhattacharya SK, Sarkar S (2007) Carbon and oxygen isotopic variations in peritidal stromatolite cycles, Paleoproterozoic Kajrahat Limestone, Vindhyan Basin of Central India. J Asian Earth Sci 29:823–831

Banerjee S, Sarkar S, Eriksson PG (2014) Palaeoenvironmental and biostratigraphic implications of microbial mat-related structures: examples from modern Gulf of Cambay and Precambrian Vindhyan basin. J Paleogeogr 3:127–144

Banerjee S, Sarkar S, Eriksson PG, Samanta P (2010) Microbially related structures in siliciclastic sediment resembling Ediacaran fossils: examples from India, ancient and modern. In: Seckbach J, Oren A (eds) Microbial Mats: Modern and Ancient Microorganisms in Stratified Systems. Springer, Berlin, pp 111–129

Bland BH (1984) Arumberia Glaessner & Walter, a review of its potential for correlation in the region of the Precambrian-Cambrian boundary. Geol Mag 121:625–633

Bose S, Chafetz HS (2011) Morphology and distribution of MISS: a comparison between modern siliciclastic and carbonate settings. In: Noffke N, Chafetz H (eds) Microbial Mats in Siliciclastic Depositional Systems Through Time. SEPM Special Publication 101, pp 3–14

Bose PK, Sarkar S, Banerjee S, Chakraborty S (2007) Mat-related features from the Vindhyan supergroup in Central India. In: Schieber J, Bose PK, Eriksson PG Banerjee S, Sarkar S, Catuneanu O, Altermann W (eds) An Atlas of Microbial Mat Features Preserved Within the Clastic Rock Record. Elsevier, pp 181–188

Bouougri E, Porada H (2002) Mat related sedimentary structures in Neoproterozoic peritidal passive margin deposits of the West African Craton. Sed Geol 153:85–106

Burne RV, Moore LS (1987) Microbialites: organosedimentary deposits of benthic microbial communities. Palaios 2:241–254

Catuneanu O (2006) Principles of sequence stratigraphy. Elsevier, Amsterdam

Chakraborty PP, Das P, Saha S, Das K, Mishra SR, Paul P (2012) Microbial mat related structures (MRS) from Mesoproterozoic Chhattisgarh and Khariar basins, Central India and their bearing on shallow marine sedimentation. Episodes 35:1–11

Choudhuri A, Sarkar S, Altermann W, Mukhopadhyay S, Bose PK (2016) Lakshanhatti stromatolite, India: biogenic or abiogenic? J Palaeogeogr 5:292–310

De C (2003) Possible organisms similar to Ediacaran forms from the Bhander Group, Vindhyan Supergroup, Late Neoproterozoic of India. J Asian Earth Sci 21:387–395

De C (2006) Ediacara fossil assemblage in the upper Vindhyans of central India and its significance. J Asian Earth Sci 27:660–683

Decho AW (1990) Microbial exopolymer secretions in ocean environments: their role(s) in food webs and marine processes. Ocean Mar Bio Ann Rev 28:73–154

Decho AW (2000) Exopolymer microdomains as a structuring agent for heterogeneity within microbial biofilms. In: Riding R, Awramik SM (eds) Microbial Sediments. Springer, Berlin, pp 9–15

Demicco RV, Hardie LA (1994) Sedimentary structures and early diagenetic features of shallow marine carbonate deposits. SEPM Atlas Series 1, Tulsa, Oklahoma

Eriksson PG, Simpson EL, Eriksson KA, Bumby AJ, Steyn GL, Sarkar S (2000) Muddy roll-up structures in siliciclastic interdune beds of the c. 1.8 Ga Waterberg Group, South Africa. Palaios 15:177–183

Eriksson PG, Sarkar S, Samanta P, Banerjee S, Porada H, Catuneanu O (2010) Paleoenvironmental context of microbial mat-related structures in siliciclastic rocks. In: Seckbach J, Oren A (eds) Microbial Mats: Modern and Ancient Microorganisms in Stratified Systems, Cellular Origin, Life in Extreme Habitats and Astrobiology, vol 14. Springer, pp 71–108

Fedonkin MA (1980) New representatives of Precambrian coelenterates on the north of the Russian platform. Paleontologičeskij Žurnal 2:7–15

Fedonkin MA (1984) Promorfologia vendckikh Radiala (Promorphology of the Vendian Radiala). Stratigrafiya I Paleontologiya revnevshego Fanerozova. Nauka, Moscow, pp 30–58 (in Russian)

Flügel E (2013) Microfacies of carbonate rocks: analysis, interpretation and application. Springer, Berlin

Gehling JG (1999) Microbial mats in terminal Proterozoic siliciclastics: Ediacaran death masks. Palaios 14:40–57

Gerdes G, Krumbein WE, Reineck HE (1994) Microbial mats as architects of sedimentary surface structures. In: Krumbein WE, Paterson DM, Stal LJ (eds) Biostabilization of Sediments. Bibliotheks und Informationssytem der Carl von Ossietzky Universität, Oldenburg, Germany, pp 165–182

Gerdes G, Klenke T, Noffke N (2000) Microbial signatures in peritidal siliciclastic sediments: a catalogue. Sedimentology 47:279–308

Gerdes G, Claes M, Dunajtschik-Piewak K, Riege H, Krumbein WE, Reineck HE (1993) Contribution of microbial mats to sedimentary surface structures. Facies 29:61–74

Glaessner MF, Walter MR (1975) New Precambrian fossils from the Arumbera Sandstone, Northern Territory, Australia. Alcheringa 1:59–69

Grazhdankin D, Gerdes G (2007) Ediacaran microbial colonies. Lethaia 40:201–210

Grotzinger JP (1990) Geochemical model for proterozoic stromatolite decline. Am J Sci 290:80–103

Hagadorn JW, Bottjer DJ (1997) Wrinkle structures: microbially mediated sedimentary structures common in subtidal siliciclastic settings at the Proterozoic-Phanerozoic transition. Geology 25:1047–1050

Hagadorn JW, Bottjer DJ (1999) Unexplored microbial worlds. Palaios 14:1–2

Kalkowsky E (1908) Oolith und Stromatolith in norddeutschen Buntsandstein. Zeitschrift der Deutschen Gesellschaft für Geowissenschaften 60:68–125

Kulkarni KG, Borker VD (1996) Occurrence of *Cochlichnus* Hitchcock in the Vindhyan Supergroup (Proterozoic) of Madhya Pradesh. J Geol Soc India 47:725–729

Lan ZW, Chen ZQ (2013) Proliferation of MISS-forming microbial mats after the late Neoproterozoic glaciations: evidence from the Kimberley region, NW Australia. Precam Res 224:529–550

Luo M, Chen ZQ, Hu S, Zhang Q, Benton MJ, Zhou C, Wen W, Huang JY (2013) Carbonate reticulated ridge structures from the lower middle Triassic of the Luoping Area, Yunnan, Southwestern China: geobiologic features and implications for exceptional preservation of the Luoping Biota. Palaios 28:541–551

McIlroy D, Walter MR (1997) A reconsideration of the biogenicity of Arumberia banksi Glaessner and Walter. Alcheringa 21:79–80

McIlroy D, Crimes TP, Pauley JC (2005) Fossils and matgrounds from the Neoproterozoic Longmyndian Supergroup, Shropshire, UK. Geol Mag 142:441–455

Menzies J (1990) Sand intaclasts in a diamicton melange, southern Niagara Peninsula, Ontario, Canada. J Quat Sci 5:189–206

Noffke N (2000) Extensive microbial mats and their influences on the erosional and depositional dynamics of a siliciclastic cold water environment (Lower Arenigian, Montagne Noir, France). Sed Geol 136:207–215

Noffke N (2007) Microbially induced sedimentary structures in Archean sandstones: a new window into early life. Gond Res 11:336–342

Noffke N (2010) Microbial Mats in Sandy Deposits From the Archean Era to Today. Springer, Heidelberg

Noffke N, Awramik SM (2013) Stromatolites and MISS-differences between relatives. GSA Today 23:4–9

Noffke N, Knoll AH, Grotzinger JP (2002) Sedimentary controls on the formation and preservation of microbial mats in siliciclastic deposits: a case study from the Upper Neoproterozoic Nama Group, Namibia. Palaios 17:533–544

Noffke N, Gerdes G, Klenke T (2003a) Benthic cyanobacteria and their influence on the sedimentary dynamics of peritidal depositional systems (siliciclastic, evaporitic salty, and evaporitic carbonatic). Earth Sci Rev 62:163–176

Noffke N, Hazen R, Nhelko N (2003b) Earth's earliest microbial mats in a siliciclastic marine environment (2.9 Ga Mozaan Group, South Africa). Geology 31:673–676

Noffke N, Gerdes G, Klenke T, Krumbein WE (1997) Biofilm impact on sedimentary structures in siliciclastic tidal flats. Courier Forschungsinst Senckenberg 201:297–305

Noffke N, Gerdes G, Klenke T, Krumbein WE (2001) Microbially induced sedimentary structure—a new category within the classification of primary sedimentary structures. J Sed Res 71:649–656

Noffke N, Christian D, Wacey D, Hazen RM (2013) Microbially induced sedimentary structures recording an ancient ecosystem in the *ca.* 3.48 billion-year-old Dresser Formation, Pilbara, Western Australia. Astrobiology 13:1103–1124

Parizot M, Eriksson PG, Aifa T, Sarkar S, Banerjee S, Catuneanu O, Altermann W, Bumby AJ, Bordy EM, Rooy JLV, Boshoff AJ (2005) Microbial mat-related crack-like sedimentary structures in the c. 2.1 Ga Magaliesberg Formation sandstones, South Africa. Precam Res 138:274–296

Pflüger F (1999) Matground structures and redox facies. Palaios 14:25–39

Pflüger F, Greese PG (1996) Microbial sand chips—a non-actualistic sedimentary structure. Sed Geol 102:263–274

Pflüger F, Sarkar S (1996) Precambrian bedding planes-bound to remain. GSA Ab Prog 28:491

Prave AR (2002) Life on land in the Proterozoic: evidence from the Torrodonian rocks of northwest Scotland. Geology 30:811–814

Reineck HE, Gerdes G, Claes M, Dunajtschik K, Riege H, Krumbein WE (1990) Microbial modification of sedimentary surface structures. In: Heling D, Rothe P, Förstner U, Stoffers P (eds) Sediments and Environmental Geochemistry. Springer, Berlin, pp 254–276

Sarkar S, Banerjee S (2007) Some unusual and/or problematic features. In: Schieber J, Bose PK, Eriksson PG, Banerjee S, Sarkar S, Catuneanu O, Altermann W (eds) An Atlas of Microbial Mat Features Preserved Within the Clastic Rock Record. Elsevier, Amsterdam, pp 145–147

Sarkar S, Bose PK (1992) Variations in Proterozoic Stromatolites over a transition from basin plain to nearshore subtidal zone. Precam Res 56:139–157

Sarkar S, Banerjee S, Bose PK (1996) Trace fossils in Mesoproterozoic Koldaha Shale, Central India and their implications. N Jb Palaont Mh 7:425–438

Sarkar S, Banerjee S, Eriksson PG (2004) Microbial mat features in sandstones illustrated. In: Eriksson PG, Altermann W, Nelson DR, Mueller WU, Catuneanu O (eds) The Precambrian Earth: Tempos and Events. Elsevier, Amsterdam, pp 673–675

Sarkar S, Banerjee S, Eriksson PG, Catuneanu O (2005) Microbial mat control on siliciclastic Precambrian sequence stratigraphic architecture: examples from India. Sed Geol 176:195–209

Sarkar S, Banerjee S, Samanta P, Jeevankumar S (2006) Microbial mat-induced sedimentary structures in siliciclastic sediments: examples from the 1.6 Ga Chorhat Sandstone, Vindhyan Supergroup, M.P., India. J Earth Sys Sci 115:49–60

Sarkar S, Bose PK, Samanta P, Sengupta P, Eriksson PG (2008) Microbial mat mediated structures in the Ediacaran Sonia Sandstone, Rajasthan, India, and their implications for Proterozoic sedimentation. Precam Res 162:248–263

Sarkar S, Samanta P, Altermann W (2011) Setulfs, modern and ancient: formative mechanism, preservation bias and palaeoenvironmental implications. Sed Geol 238:71–78

Sarkar S, Banerjee S, Samanta P, Chakraborty N, Chakraborty PP, Mukhopadhyay S, Singh AK (2014) Microbial mat records in siliciclastic rocks: examples from four Indian Proterozoic basins and their modern equivalents in Gulf of Cambay. J Asian Earth Sci 91:362–377

Sarkar S, Choudhuri A, Mandal S, Eriksson PG (2016) Microbial mat-related structures shared by both siliciclastic and carbonate formation. J Palaeogeogr 5:278–291

Sarkar S, Choudhuri A, Mandal S, Bose PK (2018) Flat pebbles and their edge-wise fabric in relation to 2-D microbial mat. Geol J. https://doi.org/10.1002/gj.3312

Schieber J (1998) Possible indicators of microbial mat deposits in shales and sandstones: examples from the Mid-Proterozoic Belt Supergroup, Montana, USA. Sed Geol 120:105–124

Schieber J (1999) Microbial mats in terrigenous clastics: the challenge of identification in the rock record. Palaios 14:3–12

Schieber J (2004) Microbial mats in the siliciclastic rock record: a summary of diagonostic features. In: Eriksson PG, Altermann W, Nelson DR, Mueller WU and Catuneanu O (eds) The Precambrian Earth: Tempos and Events, pp 663–672

Schieber J, Bose PK, Eriksson PG, Banerjee S, Sarkar S, Altermann W, Catuneanu O (eds) (2007) Atlas of microbial mat features preserved within the siliciclastic rock record. Elsevier, Amsterdam

Schopf JW (1999) Cradle of Life. Princeton University Press, New Jersey

Seilacher A (1999) Bio mat-related lifestyles in the Precambrian. Palaios 14:86–93

Shen JW, Webb GE (2004) Famennian (Upper Devonian) stromatolite reefs at Shatang, Guilin, Guangxi, South China. Sed Geol 170:63–84

Sur S, Schieber J, Banerjee S (2006) Petrographic observations suggestive of microbial mats from Rampur Shale and Bijaigarh Shale, Vindhyan basin, India. J Earth Sys Sci 115:61–66

Walter M R (1972) Stromatolites and the biostratigraphy of the Australian Precambrian and Cambrian. Spec Pap Palaeontol 11

Witkowski A (1990) Fossilization processes of the microbial mat developing in clastic sediments of the Puck Bay (Southern Baltic Sea, Poland). Acta Geol Polonica 40:1–27

Xiaoying S, Chuanheng Z, Ganqing J, Juan L Yi W, Dianbo L (2008) Microbial mats in the Mesoproterozoic carbonates of the North China platform and their potential for hydrocarbon generation. J China Univ Geosci 19: 549–566

Index

© Springer Nature Singapore Pte Ltd. 2020
S. Sarkar and S. Banerjee, *A Synthesis of Depositional Sequence of the Proterozoic Vindhyan Supergroup in Son Valley*, Springer Geology,
https://doi.org/10.1007/978-981-32-9551-3

The manufacturer's authorised representative in the EU is Springer
Nature Customer Service Centre GmbH, Europaplatz 3, 69115 Heidelberg,
Germany. If you have any concerns regarding our products, please
contact ProductSafety@springernature.com

Printed and bound by CPI Group (UK) Ltd, Croydon, CR0 4YY

28/11/2025

02007692-0005